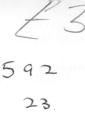

A Biology of Higher Invertebrates

Cover photo is a dorsal view of a living specimen of the marine planktonic copepod *Calanus* by electronic flash [© Douglas P. Wilson]. Though minute (2 to 4 millimeters long), calanoid copepods are probably the most numerous animals in the world (and possibly of major importance to the future nutrition of man), since they are the most abundant of the microherbivores which feed directly on the largest crop of green plants in the world: the flagellates and diatoms of the marine phytoplankton.

CURRENT CONCEPTS IN BIOLOGY
A Macmillan Series

NORMAN H. GILES, WALTER KENWORTHY, JOHN G. TORREY, Editors

A Biology of
Higher Invertebrates

W. D. Russell-Hunter

Syracuse University and The Marine Biological Laboratory

The Macmillan Company
Collier-Macmillan Limited, London

PRINTING   8910   YEAR   56789

Material reprinted from **A Biology of Lower Invertebrates,**
by W. D. Russell-Hunter, © copyright 1968 by W. D. Russell-Hunter.

Library of Congress catalog card number: 69–12171

THE MACMILLAN COMPANY
866 Third Avenue, New York, New York 10022

COLLIER-MACMILLAN CANADA, LTD., Toronto, Ontario

Printed in the United States of America

Nothing in nature stands alone. . . .—JOHN HUNTER, 1786

*. . . dans cette dépendance mutuelle des fonctions et dans ce se-
cours qu'elles se prêtent réciproquement . . . il est évident que l'har-
monie convenable entre les organes qui agissent les uns sur les autres,
est une condition nécessaire de l'existence de l'être auquel ils appartien-
nent. . . .*—GEORGES J.-L. N.-F. CUVIER, 1800

Preface

THIS BOOK—like the rest of this series—is intended for the beginning
student. It does not require any previous knowledge of the invertebrates, but assumes only background acquaintance with some aspects
of basic biology. Like its companion volume on the lower invertebrates,
it is "*a* biology" providing only *one* of the many possible surveys
of the subject. It will become obvious to the reader that much is
omitted—or given an over-narrow synoptic treatment—while a few
topics are perhaps overemphasized. Thus it seems valuable to try to
define and defend this treatment here.

About ninety per cent of described animal species (or roughly one
million living species) are "many-celled invertebrates." Even a summary catalog or directory listing of the diversity would be impossible
in two books of this size without attempting to incorporate anything of
our current understanding of the biological mechanisms which operate
within this diversity. The way out of this *impasse* adopted here results
from two natural characteristics of invertebrate diversity and two arbitrary decisions personal to the author. The first significant characteristic is that underlying the fantastic structural and functional diversity
of the invertebrates are certain *functional homologies* of considerable
extent. In other words, if comparative physiology is studied at the level
of the integration of functions in the whole animal, certain unifying
features become apparent in the ways in which each of the (possibly
thirty) animal phyla carries out and regulates its major biological
processes (see Chapter 1). A second feature of the diversity is also important. Without becoming too involved in the difficult question of
what constitutes "evolutionary success," we can be justified in regarding certain phyla as of major importance and others (the majority) as
of minor importance. As discussed in Chapter 1, in addition to using
criteria such as numbers of species in each phylum, there is a quanti-
fiable ecological sense by which designation of "minor phyla" can be

vindicated. These two books attempt to emphasize the functional homologies of major phyla at the expense of the rest.

The other limitations of the treatment are more arbitrary, but still defensible. As has been discussed many times, biological organization can be studied at several grades of complexity (e.g., at such grades as ecosystem, community, population, individual organism, tissue, cell, or molecule), and by asking questions at about four conceptual levels. These latter range from the structural-descriptive (encompassing anatomy, histology, and even some part of molecular biology on the one hand and ecological zonation on the other), through the mechanistic-physiological ("how does it work?"), and the adaptive-functional ("what is the biological value of this structural organization or group of processes: (a) to the individual animal and (b) to a population of that species?"), to the evolutionary-historical ("what is the history of this organ or process in time—how has it evolved?"). The treatment of the biology of the invertebrates here is intermediate in regard to both grade of organization and level of concept: it is concerned with whole animals considered mainly at the mechanistic-physiological and adaptive-functional levels of explanation. Such "zoological" biology (involving the study of the whole animal as an integrated functional system) may be considerably older than the presently more fashionable "molecular" biology, but *both* must be, equally, parts of the biology of the future. Even on the pragmatic level of what it "profits" man to understand, we shall need zoöcentric biologists. To put this in rather different terms, the book involves an attempt to steer between pure comparative anatomy with classification on the one hand and physiology at the biochemical level on the other. Obviously, it would be possible to devote several books of this size to a detailed account of comparative anatomy in one major phylum—say the Mollusca. Alternatively, such a book could discuss the occurrence through the invertebrate groups of a few chosen enzyme systems or blood pigments (incidentally, the data are not yet available for many enzyme systems). Neither extreme is attempted here. What *is* discussed is largely functional morphology through the various groups, involving the adaptive level of consideration of form and function in attempting to explain the way in which invertebrates of diverse groups fit their particular habitats and function in them. There is, perhaps, some overemphasis on the mechanics of locomotion and of nutrition.

The functional morphology is linked by use of those parts of systematics which are felt to be phyletically significant and by discussions of the evolution of certain characteristic functions. Other links are provided by emphasizing the existence of a few successful plans of functional organization (i.e., the existence of functional homologies within each phylum). The student should realize that both strands of continuity—the phylogenetic classification and the recognition of homologous features—are matters involving hypotheses. On the other hand, the principal content of these books—the functional morphology—is matter involving only simple observations and comparative experiments, which can be verified, or expanded, or corrected by any student, *without* elaborate apparatus but *with* access to living specimens of the invertebrate animals concerned. The use of "representative types" is deliberately avoided. Our understanding of the biology of invertebrates has suffered in the past through a pedagogical devotion to *Hydra* as "the representative coelenterate," "the earthworm" as "type annelid," *Helix* as "type mollusc," and "the crayfish" as "type crustacean." Apart from certain basic biological absurdities in this usage, the actual animal forms most employed as "types" were often (as in the above four examples) rather unrepresentative of form and function for the phylum concerned. In these books—as in teaching—the author prefers all the dangers involved in setting up a "hypothetical primitive form" as a model and then discussing the facts of diversity in a group in relation to that model to using as a starting point "the type"—the detailed study of a specialized form. The general disadvantages resulting from the use of models should be well known to every scientist, and the peculiar dangers of "archetypes" (with their values in comprehensibility) are discussed in Chapters 1, 14, and elsewhere.

An attempt has been made to allow the companion volume, *A Biology of Lower Invertebrates,* and this book to be useful even if read independently. This has involved some repetition: most of Chapter 1 is identical in the two books and, in addition, the concept of metamerism and its characteristic morphogenesis in annelid worms is discussed in both. The two volumes treat invertebrate biology largely phylum by phylum. If the significance of functional homology within each phylum (see Chapter 1) is acceptable to the reader, then it will be realized that this treatment is followed, not because it happens to have been the classical pattern for texts on invertebrate anatomy, but because the author believes it to be conceptually in line with current thinking about comparative physiology, and thus in accordance with the general title of the series.

Again, I must thank all who helped with illustrations. My special gratitude to Dr. Douglas P. Wilson of the Plymouth Laboratory for the inimitable photographs of marine invertebrates, which grace both books, commemorates more than twenty years of scientific exchanges. The skill of Peter Loewer in translating my colored cartoons into an integrated series of line drawings has added much of value to both books, and I am most grateful. It is a pleasure to thank Professor John M. Anderson of Cornell; Professor John H. Welsh of Harvard; my research associates—Robert J. Avolizi and Jack S. Mattice; and

my wife—Myra Russell-Hunter; all have contributed helpful sugges-
tions on reading the original manuscript of this book. Obviously, I am
solely responsible for the errors that remain and uniquely to blame for
the uneven emphasis on certain topics and groups.

W. D. R-H.

Contents

A Biology of Higher Invertebrates

Introduction: Functional Homologies and Phyla

THERE ARE PROBABLY NEARLY two million living species of animals, approximately half of which have been formally described. Of these, approximately 8 per cent are protistan and approximately 3.5 per cent are vertebrate, leaving about 90 per cent of described animal species as "invertebrates." A synoptic view of the thirty phyla of many-celled animals is provided in Table 1·1 (p. 5). It should be noted that the numbers of species listed are approximate and in some cases compromise figures between widely differing estimates. Further, estimates of the total number of extinct species of animals are mostly around seven times the number presently alive, and the proportion of these which could be termed "many-celled invertebrates" is probably about the same. In other words, there may have been about thirteen million invertebrate species.

Animal Phyla

As is well known, species can be delimited on the basis of a "biological species definition" which is based on the natural (nonarbitrary) criterion of reproductive isolation. The practical application of this criterion in taxonomic discrimination is usually difficult, and almost all described species involve arbitrary separations based on subjective inferences about genetic groupings, which are themselves derived from "hard" data on morphological groupings and on their geographical and ecological distribution. All taxa above the species level are admittedly ideal, arbitrary categories rather than natural categories. However, it has often been recognized that, among the higher taxa of systemics, a phylum—particularly a phylum of metazoans—is normally less subjective than such a taxon as class or order. Although a logically defined grouping like these intermediate taxa, a phylum may approach the objective reality of a species more closely.

The aspect of this most often stressed is that the thirty animal phyla are characterized by a unity of basic structural pattern in each. In other words, though the members of a phylum may vary in external features, comparative anatomical studies reveal that they form an assemblage, all constructed on the same ground plan with certain essential groupings of structural units. Much less emphasis has been placed on the considerable functional unity which exists within each phylum. If comparative physiology is studied at the level of the organization of the whole individual animal, certain unifying features become apparent in the ways in which each animal phylum carries out and regulates its major biological processes. In other words, underlying the fantastic structural and functional diversity of the invertebrates are certain functional homologies of considerable extent. Throughout this book and its companion volume on "lower" invertebrates, an attempt is made to stress these homologies. The concept of homology—whether applied to structures or functions—involves the hypothesis of (theoretically) traceable derivation from, and origin in, some *common* anatomical or physiological precursor in a *common* ancestral animal. Thus, like all other concepts involving phylogeny (the circumstantially deduced history of genealogical descent), including "archetypes as models of ancestors," homologies are hypothetical. Perhaps it is safest to emphasize the pragmatic value of functional homologies, in that they can allow human comprehension of the diversity of form and function in the invertebrates. However, some assessment of the implicit biological meaning of structural and functional homologies is in order.

Within each phylum of animals, the common anatomical ground plan implies a unique network of relationships between the groups of structural units which compose it. Some of this pattern of relationships is "stamped in" early in development by the need for certain early sequences of differentiation to precede the more elaborate processes of morphogenesis of organs which must follow. The significance of this necessity for stereotyped pattern was outlined by O. Hertwig and R. Hertwig at the end of the nineteenth century, and since has intrigued (and sometimes misled) many developmental biologists. (See the volumes by C. H. Waddington and John W. Saunders in this series.) Another significant feature of the pattern of the interrelationships is functional. All animals must be efficient machines, in which the whole concert of organs and functions operates in an integrated fashion. Further, all ancestors of animals must have been working, efficient machines with similar functional integration. The need for physiological interaction in the way each major biological process is carried out and regulated has imposed pattern. The thirty "patterns" that we call phyla *do* work as efficient machines. Natural selection has not only tended to maintain each stereotyped network of anatomical relationships (and eliminated almost all random variants), but it has promoted increasing functional interdependence in a pattern of whole animal function characteristic for each phylum.

Perhaps this somewhat theoretical discussion of the nature of functional homologies can be elucidated by a few specific negative examples. Characteristic hollow skin structures termed tube-feet or podia are found throughout the phylum Echinodermata. Although employed in a variety of ways—for the manipulation of food or protective materials, for respiration, for sensory purposes, and for locomotion—it is accepted, on the basis of extensive morphological evidence, that podia are structurally and functionally homologous in about six thousand species of living echinoderms (and probably in at least another forty thousand extinct forms). They appear as numerous delicate external projections from the surface of the animal, each having: intrinsic muscles for withdrawal and for postural "pointing," a hydraulic mechanism for protraction depending on the so-called water vascular system within the animal, a characteristic pattern of internal ciliation for fluid circulation, and individual innervation which allows either local reflex responses or movements integrated throughout the whole animal. Podia, or tube-feet, are efficient structures for certain kinds of slow locomotion and manipulation and could seemingly be used by many aquatic animals belonging to other phyla. *Only echinoderms have them.* Within the echinoderm body, the podia are part of an integrated functional system. Mechanically, they could not possibly function unless they were supported by the unique skeleton of dermal ossicles which give the phylum its name. Although the details of protraction vary in different echinoderms, all podia are extended hydraulically by pressure of fluids within the water vascular system, a subdivision of the coelom. Thus, they depend on the peculiar interrelationships of development and adult function of the coelomic spaces which are unique to the phylum Echinodermata.

Other similar negative examples can be quoted. Many animals, other than coelenterates, could use the characteristic stinging-cells (nematocysts) for defense or offense. *No others make and use them.*

A characteristic form of gill, the ctenidium, is found throughout most of the phylum Mollusca and is inferred (on evidence that leaves little room for any doubt) to be structurally and functionally homologous throughout those animals where it occurs (i.e., in perhaps seventy-five thousand molluscan species; see *A Biology of Lower Invertebrates,* hereinafter referred to as BLI). It is functionally a most competent organ, with patterns of ciliated epithelia and blood vessels arranged to create a highly efficient counterflow system for oxygen exchange between blood and water, characteristic cleansing mechanisms (both ciliary and muscular), and typical mechanical arrangements of supporting skeletal elements. There are many aquatic animals belonging to other phyla which seemingly could make good use of a ctenidium. *No nonmolluscan animal has one.* Within the molluscs, the ctenidia are part of an integrated functional system: the heart and other blood vessels, certain glands and sense-organs, the external openings of genital and renal systems, and the posterior part of the alimentary canal are all structurally and functionally stereotyped in their relationships to the ctenidia. Many more examples of this sort could be set out.

A related topic is the use of "hypothetical primitive forms" or "archetypes" as models. Biologists will never know with certainty the characteristics of the ancestors of *any* stocks of animals. All modern scientists should be well aware of the general pitfalls—as well as the pragmatic values—of the use of models. The peculiar dangers in evolutionary discussions of setting up an "archetype" seem to result from assembling together in the unfortunate hypothetical animal a group of incompatible structures, all thought to be "primitive" within the stock. The author feels that much of these can be avoided if, when a hypothetical ancestral type is constructed, an attempt is made to create a working archetype—one in which the concert of organs and functions could operate as a whole, in an integrated functional plan, as in all living organisms. It is perhaps significant that discerning functional homologies and setting up archetypes is much *more* difficult in the phyla of *less complex* animals. Significant functional unity is more apparent in the Mollusca, the Arthropoda, or the Echinodermata than it is in phyla with less complexity of structure and function such as the Cnidaria or the Platyhelminthes. It seems that the network of functional relationships is less closely integrated in these "lower" groups, and thus the anatomical and physiological pattern not so stereotyped. "Experimental variants" of pattern can be thought of as having a better chance of survival in a group with less complexity of interdependence in its basic archetype.

Major and Minor Phyla

If we accept the logically defined grouping of many-celled animals into the thirty phyla of Table 1·1 as approaching some objective reality of common descent and interrelationship, then it is immediately obvious from the species numbers that some "patterns" of animals are more successful than others. Usually the designation of the major

TABLE 1·1

Approximate Numbers of Living Species of Many-celled Animals [1]

PHYLUM	SPECIES	
Cnidaria	11,000	Jellyfish, Sea-anemones, Corals, Hydroids, etc.
Ctenophora	80	Comb-jellies, Sea-gooseberries
Porifera	4,200	Sponges
Mesozoa	50	......
Platyhelminthes	15,000	Flatworms: planarians, flukes, tape-worms
Rhynchocoela (= Nemertea)	600	Ribbon-worms
Entoprocta	60	Moss-animals[3]
Rotifera	1,500	Wheel-animalcules
Gastrotricha	150	......
Echinorhyncha	100	......
Nematomorpha	250	Horsehair-worms
Acanthocephala	300	Spineheaded-worms
Nemathelminthes	80,000[2]	Roundworms: free-living and parasitic
Mollusca	110,000	Snails, Clams, Octopus, etc.
Annelida	8,800	Segmented worms
Arthropoda	>800,000	Insects, Crustaceans, Spiders
Onychophora	80	Walking-worms
Tardigrada	170	Water-bears
Linguatulida	60	Tongue-worms
Echiuroidea	80	......
Ectoprocta	4,000	Moss-animals[3]
Priapulida	5	......
Phoronida	15	......
Brachiopoda	310[4]	Lampshells
Sipunculoidea	275	......
Chaetognatha	60	Arrow-worms
Pogonophora	45	......
Echinodermata	6,000	Starfish, Sea-urchins, Sea-cucumbers, etc.
Hemichordata	100	Acorn-worms
Chordata	45,000	Vertebrates, etc.
(includes "Invertebrate Chordata")	2,100	Lancelets, Sea-squirts

Notes:
[1] The clear area shows the phyla discussed in this volume.
[2] This is a compromise figure, between widely differing estimates; see text.
[3] The common name "moss-animals" is applied to two distinct phyla.
[4] Plus at least 12,000 described fossil species.

phyla is based on the number of individuals, the number of species, or the number of "successful, dominant" species encompassed in the groups, or on (and this is almost impossible to quantify) the relative "phyletic importance" of the groups. On all such counts, four phylum-groups at least are clearly major: arthropods, chordates, molluscs, and nematodes. More than twelve phyla of invertebrates are equally clearly minor. It is less often recognized that such designation can also be justified ecologically, involving a relatively simple aspect of community ecology which has become more and more obvious to the present author over the last fifteen years. Briefly, it is that the bulk of the animal biomass in the great majority of natural communities is made up of individuals belonging to a few major phyla of animals, and that the minor phyla make up only a tiny fraction of the animal tissue alive in the world today. To put it another way, of the solar energy first incorporated into green plants, a disproportionately large share flows through (or is utilized by) representatives of such major phyla as the arthropods, whereas the energy-flow through the representatives of a phylum such as the Entoprocta is normally several orders of magnitude smaller. Thus assessed, it seems justifiable to refer to the phylum Entoprocta as "relatively unsuccessful" in the competition for the *finite* amount of organic energy originating in the green plants of the world, and to refer to it as a minor phylum. There are probably thirty phyla of many-celled animals. Utilizing a combination of the above criteria, we can regard eight as major phyla, perhaps three others as important, and the rest as minor phyla of little phyletic or ecological significance. The minor phyla are not treated extensively here.

The phyla discussed in this volume are indicated in Table 1·1, and all sixteen show the "triploblastic coelomate" basic plan (see Figure 2·1), found in the great majority of successful groups (including Mollusca, Annelida, Arthropoda, Echinodermata, and Chordata). As discussed in BLI (pp. 7 and 66), the triploblastic phyla have their intermediate layers formed of true mesoderm (or entomesoderm, the great bulk originating as an initially epithelial mass of tissue or as wandering cells derived from the endoderm, and relatively little as wandering amoeboid cells derived singly from the ectodermal epithelium), but also—and perhaps of greater functional significance—they show an "organ grade" of interdependence of parts. The coelomate condition implies the possession (at least developmentally) of a body-cavity within the mesoderm. The internal organs are slung on mesodermal mesenteries and there is a mesodermal covering to the gut-tube endoderm as well as a mesodermal lining to the outer body wall (see Figure 2·1, and BLI pp. 85–86). The two major patterns of coelom formation, and their unfortunate inefficacy when used for phyletic distinctions, are discussed in Chapter 9 of this book. Thus, in respect of overall structural complexity and extent of functional integration in each phylum-pattern, the use of the terms "lower invertebrates" and "higher invertebrates" in these two books can almost be justified. Paradoxically, however, the most highly organized invertebrate animals—the large, fast-moving, "brainy" cephalopods (including *Octopus* and its allies)—are built on the molluscan pattern and thus fall in the other volume (BLI) on the "lower invertebrates." The phylum Annelida is discussed both in this volume and in BLI, a fuller account of archetypic functional organization (apart from locomotion) and of worm diversity being given here (Chapters 2 and 3). Annelid locomotion is discussed in BLI (Chapter 8, pp. 104–111), where it can be related to the locomotory mechanisms found in other wormlike animals.

More than half of this volume (Chapters 2–8) is devoted to the major and minor groups of metamerically segmented animals (see Table 2·1). The rest is concerned with several minor phyla, including those using a ciliated tentacular crown (or lophophore) for food collection, with the Echinodermata, and with the invertebrate Chordata and their allies. In order to deal with these sixteen phyla (see Table 1·1), a somewhat arbitrary treatment of the enormous phylum Arthropoda has been evolved. The biology of one major class, the Crustacea, is treated more extensively, largely because the occurrence of relatively primitive marine forms within it makes crustacean functional morphology a more rewarding study. By convention, many texts on invertebrates omit the insects entirely. In a compromise treatment adopted here, insect and arachnid diversity are surveyed only synoptically, but certain aspects of insect function which are of major significance to comparative physiology—including respiration, excretion, molt-cycle control, land locomotion, and flight—are discussed more extensively (see Chapters 4 and 7).

2

Segmental Organization: Annelida-Arthropoda

No TWO PHYLA of the more complex animals share more features of structural and developmental organization than the annelids and arthropods. The phylum Annelida encompasses the segmented worms, of the sea, of fresh waters, and of terrestrial soils. The phylum Arthropoda, the enormous phylum of the most successful invertebrate stocks (see Table 1·1), comprises the insects, crustaceans, spiders, and other groups, all with a jointed, chitinous procuticle, serving both as a protective armor and an exoskeleton.

The structural homologies *common* to the two groups are mainly concerned with early development and with the nature of metameric segmentation. The patterns of functional interdependence, while somewhat stereotyped within each group, clearly separate the two phyla. Functional integration in the efficient machines which are annelid worms differs from that in arthropodan machines, and the differences are chiefly those associated with the nature of the integument: differences in the mechanics of growth, of respiration, of excretion, and of locomotion. Considering, for example, only the last aspect of physiology in the two metamerically segmented groups, the distinction is clear-cut. Annelid worms are soft-bodied animals, using longitudinal and circular muscles in the body wall for locomotion, the forces generated by contractions being transmitted by pressure changes in the worm's body-fluids. Thus, the efficient propulsion of an annelid through its environment involves antagonistic sets of muscles acting on each other by way of a hydrostatic skeleton and producing either undulatory movements or peristaltic waves of alternate elongations and bulgings which deform the entire "soft" body of the worm. In contrast, arthropods have a rigid exoskeleton and, eponymously, jointed limbs with internal muscles arranged antagonistically as flexors and extensors of each joint. The arthropod is propelled through the environmental medium, or over the substrate, by the movements of the interacting systems of levers which are its jointed appendages. The differences in the mechanics of locomotion in the two phyla obviously involve the structural arrangements and the functioning sequences of the muscles and of their motor innervation. Much more, however, is involved through the necessary functional integration within each type of animal machine. Not merely muscles and motor nerves, but the integrative parts of the nervous system, the sensory receptors—both external and internal, the circulatory system and heart, and even the functional organization of the alimentary canal must differ along with the pattern of locomotion. Similar interdependence would characterize a consideration of the different patterns of respiration in the two phyla.

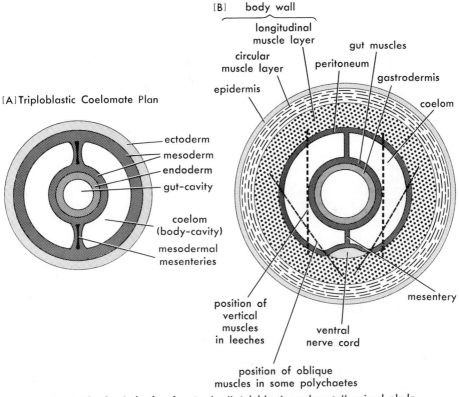

Figure 2·1. A: The basic body plan in the "triploblastic coelomate" animal phyla, which include Mollusca, Annelida, Arthropoda, and Chordata. (Compare with BLI, Figure 6·3.) B: Cross section through an archetypic annelid, showing the layers of the body wall and the positions of additional musculature in some polychaetes and leeches.

Six terms of morphology which can be applied to both Annelida and Arthropoda include: triploblastic, coelomate, bilaterally symmetrical,

TABLE 2·1

Outline Classification of the Metameric Coelomates

Phylum ANNELIDA	Class Archiannelida[1,5]
	Class Polychaeta
	Class Myzostomaria[1]
	Class Oligochaeta
	Class Hirudinea
Phylum ECHIUROIDEA[1]	
Phylum ONYCHOPHORA[1]	
Phylum ARTHROPODA	Class Trilobita[2]
	Class Merostomata[1]
	Class **CRUSTACEA**[3]
	Class **ARACHNIDA**[3]
	Class Pycnogonida[1]
	Class **INSECTA**[3]
	Class Pauropoda[1,4]
	Class Chilopoda[4]
	Class Diplopoda[4]
	Class Symphyla[1,4]
Phylum TARDIGRADA[1]	
Phylum LINGUATULIDA[1]	

Notes:
[1] Minor groups with relatively few living species.
[2] No living species.
[3] The three major "successful" groups of arthropods.
[4] The four groups commonly termed "centipedes and millipedes" or "myriapods."
[5] This minor group should probably be suppressed; see pp. 22–23 and 35.

with a tubular gut running from mouth to anus, with a ventral nerve cord, and with metameric segmentation. The implications of the first three characteristics are discussed elsewhere (see Figure 2·1, p. 6, p. 132, and BLI, p. 85). The functional significance of a tubular gut running from a subterminal mouth to a terminal anus through an elongate body is, of course, that it allows for one-way traffic of food material past a series of specialized tissues for trituration, secretion, digestion, and absorption. In other words, annelid worms and arthropods, although differently organized, have a functional alimentary layout similar to that in the gut of vertebrates, and, like them, capable of forming a dis-assembly-line for food materials. The last diagnostic character—that of metameric segmentation—implies that the body has its muscles, nerves, and internal organs subdivided in sets forming the metameric segments.

The essence of metamerism is the serial succession of segments, each containing unit subdivisions of the several organ systems. Not only the skin, the muscles, and nerves of the body wall are involved, but also, in more primitive segmented forms, such internal organs as those of circulation, excretion, and reproduction. Functionally, this type of organization is important in two ways: first, during development, in the nature of morphogenesis, and secondly, with regard to the muscle systems and hydrostatic skeleton used in locomotion in soft-bodied forms. In the truly segmented animals such as annelids and arthropods, metamerism appears very early in embryonic development. Indeed, in many, as soon as mesoderm is formed, it is organized into mesoblastic somites. Although many of the obviously segmental features are ectodermal in origin, including the parapodia in worms, the limbs in arthropods, and the segmental ganglia of the nervous system in both, segmentation appears to be basically a phenomenon of mesodermal organization and proceeds from the inside of the embryo outward. It is important to note that, in the more highly evolved worms and arthropods, the exact numbers of somites or segmental structures in the adult can often be difficult to determine, but demarcation is always clearer in the embryo or in the larval stages. Later, fusion, loss, or modification of somites may obscure the regularity of repetition of the several organ systems, or of the serial succession of segments themselves. It is in the least specialized annelid worms—presumably more primitive—that we find adults showing closest adherence to a serial pattern of identical units.

The second significant feature of metamerism in the physiology of soft-bodied segmented animals is its importance in locomotion. It is possible for acoelomate, nonsegmented animals to use body-wall musculature arranged in longitudinal and circular elements to carry out various forms of locomotion (see BLI, p. 82, on the locomotion of ribbon-worms). However, the evolution of a coelomic cavity has allowed a hydrostatic skeleton to be used in arranging the antagonistic action of these, so that faster and more powerful forces may be used to propel the coelomate animal through its environment. Further, the evolution of metameric segmentation, including the subdivision of the muscular units into a repeated series, and the subdivision by septa of the coelomic cavity itself, has allowed not only even greater locomotory efficiency, but considerably greater capacity for local changes of shape along the elongate body. The significance of this is immense, not only in burrowing, but in all other types of locomotion. Compared with a ribbon-worm, an annelid worm (coelomate and segmented) is capable of faster and more efficient crawling and burrowing, and of more powerful and more sophisticated responses to environmental dangers or to predators.

Archetypic Development: Annelids

In the most primitive marine annelid worms, the sperms or eggs produced by the ripe gonads either rupture through the body wall itself, or pass out through the segmental excretory organs, the nephridia or coelomoducts, to meet in an external fertilization in the sea. The zygote thus formed undergoes a peculiar pattern of division known as spiral cleavage (see p. 131, and John W. Saunders' *Animal Morphogenesis* in this series) to form a blastula and then, by invagination of a gut pouch, a ciliated gastrula. In a relatively primitive annelid, such as the genus *Polygordius,* this develops into the ciliated larva known as a trochosphere, which then drifts in the plankton. As shown in Figure 2·2, the trochosphere has two ciliated bands, one above the mouth and one below the mouth, which are used both for locomotion and for feeding. There is also a characteristic apical sense-organ. Subsequent development of this larva takes place largely in the area around the anus. It is in this region that metameric segments are first divided off (see Figure 2·2B). The first few segments may appear more or less simultaneously, or they may even appear to be budded off in sequence from the main body of the ciliated larva. However, the establishment of the main budding zone in the segment immediately in front of the anus (see Figure 2·2C) is soon set up, and thereafter the segments form in an anterioposterior sequence so that the segments in front are older than the segments behind, the youngest segments always being the penultimate ones immediately in front of the anal somite, which may or may not be specially modified. In subsequent larval development the segmented portion of the body grows at the expense of the original trochosphere part and the little wormlike larva develops.

At about this stage, the young worm settles out of the plankton to the sea bottom and takes up a more or less adult way of life. At this stage (see Figure 2·2C) the two head somites of the adult worm, the prostomium and the peristomium, are clearly seen to be derived from the structures of the original trochosphere: the apical sense-organ which gives rise to the prostomium with its contained supraoesophageal ganglia (or brain) and the rest of the original larva which gives rise to the peristomial segment around the mouth. The remainder of the adult worm consists of the metameric segments first divided off in the planktonic larva. If such developmental stages are examined more closely, it is seen that each segment, when it is originally budded off, consists initially of a solid-walled tube of three layers (Figure 2·2D). Paired cavities appear as splits in the mesoderm in each segment, and these form the segmentally divided pairs of coelomic cavities. As shown in the figure, these cavities enlarge until, in each segment, they are separated only by the dorsal and ventral mesenteric slings of the gut

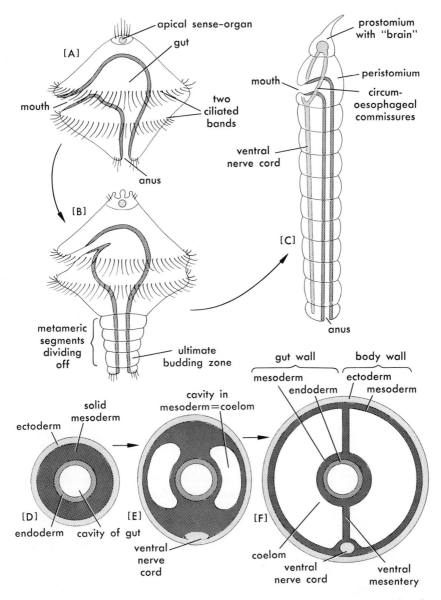

Figure 2·2. A to **C: Archetypic development of metameric segmentation in annelids. A:** Planktonic trochosphere larva of *Polygordius*. **B:** Later planktonic larva with metameric segments being budded off. **C:** Still later stage of juvenile segmented worm, about to settle out from plankton. **D** to **F:** Development of mesodermal structures and of coelom, as shown in cross sections through successive ages of budded segments.

13

and by similar thin, double septal walls between adjacent segments. Thus, in a young developing worm of 20 segments, although about 75 per cent of its volume is fluid-filled coelomic space, this space is not continuous but is subdivided transversely between each segment by a thin septum and vertically within each segment by the dorsal and ventral mesenteries, so that there is a series of forty fluid-filled coelomic cavities in a 20-segment worm.

The budding of a series of sets of segmental organs from the budding zone immediately in front of the anal segment in developing worms is paralleled by the capacity of many worms to regenerate lost segments. This will be discussed in the next chapter.

Archetypic Development: Arthropods

Most arthropods have eggs containing massive amounts of yolk. This affects early embryology, and cleavage is usually relatively superficial. A form of spiral cleavage has been reported in a few arthropods, but, even where it occurs, the pattern of mosaic development which follows differs significantly from that of annelids (see p. 132). Primitive aspects of early larval development are bound to be similarly obscured in the terrestrial habitat of most insects and arachnids. Thus it is among the largely aquatic crustaceans that a larval pattern occurs which shows acceptably archetypic features.

In some representatives of all the many diverse groups in the class Crustacea (see Table 5·1), the larva which first hatches from the egg is a nauplius. This is the simplest crustacean larva, and the least differentiated of all arthropodan larvae. The nauplius has three pairs of jointed limbs which are usually known by the names of the adult appendages into which they develop, though, as shown in Figures 2·3 and 5·6, the naupliar limbs are not specialized as head appendages but serve principally for locomotion and also for feeding. There are usually several molt stages during which the young crustacean grows but retains the nauplius form (see Chapter 4 for the physiology of the molting process). In external appearance, the nauplius is a 3-segmented larva with three pairs of appendages, but internally there is, in some cases, evidence that embryologically there are 4 segments involved. The first pair of appendages (corresponding to the adult first antennae) are always uniramous, while the other two pairs (corresponding to the adult second antennae and mandibles) are biramous and have stout bristles in their basal parts forming gnathobases (see p. 45).

Then at one molt, a fourth pair of appendages appears posteriorly (corresponding to the first maxillae of the adult crustacean), and the larva becomes a metanauplius. Although the nauplius shown in Figure 2·3A is that of a copepod, nauplii of many diverse groups of crustaceans are closely similar in form. It is only after the metanauplius stage that the different group characteristics begin to appear in the larval development (see also Figure 5·6). The later addition of more segments posteriorly is best seen in its simplest pattern in a relatively primitive form of branchiopod crustacean like *Artemia*. Newly added segments may not develop limbs until a subsequent molt. A later metanauplius (see Figure 2·3B), and an even later 20-segmented larva of a cephalocarid (see Figure 2·3C), serve to show the process of metameric addition. The process in copepods is illustrated in Figure 5·6.

It should be noted that, once again, the segments form in an anterioposterior sequence so that the segments in front are older than the segments behind. Thus, in the developing arthropod, as well as in the developing annelid worm, the youngest segments are always the penultimate ones, the main zone for the development of new metameric segments being immediately in front of the anus (Figure 2·3). In a number of crustaceans, the development of the paired limbs lags behind that of the segments which bear them, so that during the process of addition there may be, from anterior to posterior, segments with fully formed paired appendages, then well-formed segments without appendages, then poorly-differentiated segments, and then the anal parts. This can be seen most clearly in forms like anostracans and cephalocarids, which relatively undifferentiated and probably primitive forms show slow addition at each molt-stage (see Figure 2·3). Even in some other, more highly specialized, crustaceans, there is often a stage late in the larval development when there is a region in front of the anus which is relatively undifferentiated and shows incomplete segmentation. For example, this occurs in the pre-adult copepodite stages of copepods (see pp. 71–72).

It is worth noting that it is extremely easy to rear the larval stages of branchiopod crustaceans like *Artemia* from their desiccation-resistant "eggs" in a beaker of salt water without elaborate apparatus. The processes of locomotion and of feeding in nauplius and metanauplius larvae are easily studied under low magnifications, and more detailed and sequential observations can yield a real understanding of many aspects of arthropodan physiology.

The most anterior of the head structures of the adult crustacean are clearly seen to be derived from the segments and appendages of the original nauplius. The adult mouth is somewhat more posterior, lying between the mandibles, and in all but the most primitive crustaceans the adult head is formed of the first 5 appendage-bearing segments of the larval series (that is, the head is equivalent to the larva at one appendage past the first metanauplius stage). The remainder of the body of the adult crustacean consists of the metameric segments and their appendages which were first divided off during the larval development. Further features of crustacean larvae will be discussed for each

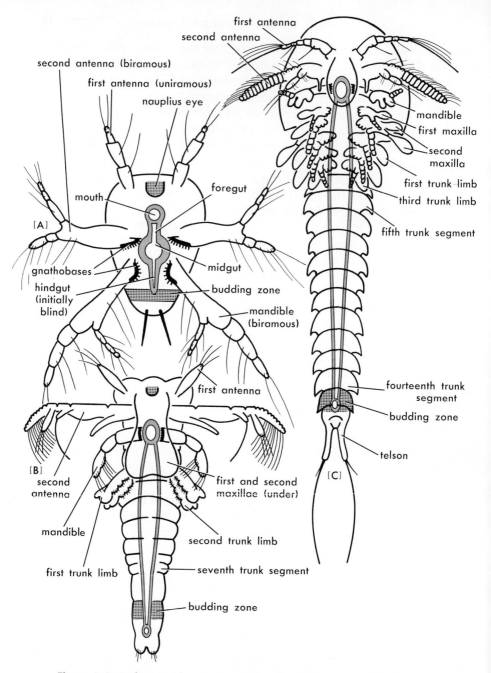

Figure 2·3. Archetypic development of metameric segmentation in crustaceans.
A: Generalized nauplius larva (basically of a cyclopoid copepod) with 3 overt segments bearing three pairs of appendages. Although these will become specialized head appendages—antennae and mandibles in the adult—they are unspecialized and used both for propulsion and food-gathering in this larva. **B:** A later metanauplius (actually of the anostracan *Artemia*) with 12 segments and seven pairs of appendages

major group later. It is of some significance, that within the likely stem group of the Arthropoda, the fossil class Trilobita, the sequence of larval stages is known in some detail, and appears to have been closely similar to that just described for *Artemia,* although the first hatched larva, the protaspis, was of 5 segments. Additional segments in trilobite development were added, once again, from a budding zone in front of the anus. To recapitulate, in spite of the major structural and functional differences between the annelids and the arthropods, the developmental processes by which segments are formed and added—the morphogenetic processes of metamerism—are closely similar in the two phyla.

Tagmatization

As mentioned earlier, in more highly evolved worms and arthropods, the adult pattern and numbers of somites or segmental structures can often be difficult to determine. Some part of this difficulty arises from the organization of tagmata (singular tagma). In relatively unspecialized annelid worms, and in some elongate arthropods like centipedes, the pattern of serial succession of identical units is most clearly seen. In these forms the structures and functions of all segments are almost uniform, all serial organs being equally developed in all segments, and many functions being carried out by every segment. In more highly organized forms we find groups of segments—the tagmata—structurally marked off from other groups and specialized to perform certain functions for the whole organism. The basis of the regional anatomical differentiation of a series of tagmata is the physiological division of labor among them. The results of tagmatization during development are most often discussed in relation to the arthropods, and it is not usually realized that the same process takes place in annelid worms to some extent. Indeed, the concept in its simplest form, of regional specialization and interdependence of regions or tagmata, is perhaps easiest to understand where it is least developed: in a series of marine bristle-worms (Annelida, class Polychaeta, see p. 21).

---

including two pairs of trunk limbs. **C:** A still later metanauplius (actually of the cephalocarid *Hutchinsoniella*) with 20 segments and eight pairs of appendages including three pairs of trunk limbs. (This larva corresponds to the stage 8 or stage 9 "nauplius" in the studies of cephalocarid development by Howard L. Sanders and Meredith L. Jones.) Note the mode of addition of segments and of paired appendages and compare with the annelid pattern of metameric development illustrated in Figure 2·2. [**C:** Modified from Howard L. Sanders in *Memoirs Conn. Acad. Arts Sciences, 15:* 1–80, 1963, with most of the setation and slighter spines omitted for clarity.]

A primitive polychaete like *Nereis* (see Figure 2·4A) can be considered as showing no tagmatization. Such a worm is a long series of like segments, uniform in structure and in their functional contribution to the whole *Nereis*-machine. A simple case where tagmata are marked off would be that of the burrowing lugworm, *Arenicola* (see p. 23). In this, the prostomium, peristomium, and first body segment have a number of features in common and can be considered to form the first tagma of 3 segments. They are followed by the next 6 body segments which bear only setae externally, and which, internally have mixonephridia as excretory organs. The third tagma consists of 13 segments which bear externally both setae and gills and which internally lack mixonephridia. Then finally, there is a fourth tagma of 21 posterior segments which do not have setae or gills or mixonephridia. Thus, for *Arenicola,* we have a total of 43 segments arranged in four tagmata. At this simple level of specialization, the interdependence is obvious. The 13 segments of the third tagma carry out most of the respiratory exchange for the whole worm, while the 6 segments of the second tagma are the segments solely responsible for excretion and water control. More complex cases of tagmatization occur among polychaete worms. The tube-dwelling worm *Chaetopterus* (see Figure 2·4B) lives in a secreted membranous tube and is a suspension-feeder utilizing mucous filtration. The anatomical and functional details need not concern us now but the 43 segments in *Chaetopterus* fall into seven clearly distinguishable tagmata.

In the great majority of arthropods the metameric segments are grouped into clearly defined tagmata. In fact, each major group of the arthropods has a nearly diagnostic pattern of tagmatization which is retained even in the most diversely adapted forms within each group. The insects show three tagmata: head, thorax, and abdomen; with a stereotyped pattern of segments involved. The insect head consists embryologically of 6 segments, the thorax consists of 3 segments which bear the only adult walking legs (hence the alternative name Hexapoda), and the abdomen has typically 11 segments without locomotory appendages. The crustaceans also show threefold tagmatization, though the segments involved do not correspond to those in the insects. The crustacean head has 5 obvious segments (but 6 embryonically) all of which bear appendages, including the two pairs of antennae anterior to the mandible and adult mouth. In the most typical crustacean pattern there follows an 8-segmented thorax, all segments bearing limbs, and a 6-segmented abdomen, all segments of which can bear limbs. The class Arachnida (spiders, mites and their allies) have the body divided into a prosoma and an abdomen or opisthosoma. Fusion of segments has occurred in both tagmata. The prosoma seems solid and nonmetameric in most adult arachnids though it consists developmentally of 3 "head" segments, two bearing appendages, and the 4 segments which bear the eight legs. In most arachnids the opisthosoma is a single tagma of 12 completely fused segments, but in primitive arachnids, like scorpions, there are two tagmata: a preabdomen of 7 segments and a postabdomen of 5 segments. The body of the fossil trilobites was divided into three tagmata: a cephalon, a thorax or trunk, consisting of a varying number of separately articulated segments, and a posterior pygidium. (The name, however, refers to a threefold lateral division by a pair of anterioposterior furrows which involves all segments.)

To sum up, only in unspecialized and presumably more primitive forms in both phyla, do we find adult animals with a strict metameric pattern of serial succession of identical units. In more highly evolved

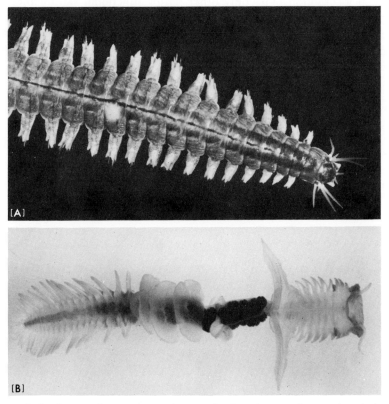

Figure 2·4. Contrasting segmental organization in two forms of polychaete worms. **A:** Anterior end of *Nereis diversicolor*, with a long series of uniform segments behind the head, and no groups of segments (or tagmata) modified for special functions. **B:** A specimen of *Chaetopterus variopedatus* removed from its tube. There is a high degree of tagmatization involving structural and functional specialization of seven groups (tagmata) of segments. [Photos of living worms both © Douglas P. Wilson.]

forms, fusion, loss, or modification of somites may obscure the regularity of pattern, as do specialized groupings of segments or tagmata. Tagmatization is in general more extensive in arthropods than in annelids. This is understandable because in the organization of soft-bodied forms like annelids, the value of metameric segmentation to locomotion is higher. With the acquisition of an arthropodan integument as an exoskeleton, segmented animals no longer required fluid-filled body cavities for the hydrostatic purpose of muscle antagonism. Thus in arthropods, tagmatization can be much more comprehensive, involving all organ systems and functions in the regional specialization of segments. However, it remains probable that the original evolution of metamerism, and indeed of the coelom or body-cavity, was governed by its value with regard to the muscle systems and hydrostatic skeleton used in locomotion. The only other major functional significance lies in the nature of segmental morphogenesis during development. Taking the hypothesis a little further, it might be concluded that metameric segmentation could only have arisen in a relatively elongate soft-bodied animal with a body-cavity.

Annelid Diversity

THE SEGMENTED SOFT-BODIED WORMS—of the sea, of fresh waters, and of terrestrial soils—make up the phylum Annelida. Encompassing nearly nine thousand known species, the phylum is also a major one in terms of the ecological measure exemplified earlier. The diagnostic features of the group have already been discussed. Functionally, the most significant is metamerism: there is a longitudinal division of the body wall and of the body-cavity into a series of segments, each containing unit subdivisions of the several organ systems.

The phylum is subdivided into three major and two very minor classes. The most numerous and most diverse group encompasses the bristle-worms, the almost entirely marine class Polychaeta, comprising more than five thousand three hundred species. Polychaetes are mostly dioecious, and spawn their eggs and sperms into the sea so that fertilization is external and early development involves ciliated planktonic larvae, like those described for *Polygordius*. Among the bristle-worms, there are several unspecialized families which show little or no tagmatization, including some which are probably closest to the most primitive annelid pattern. These are families of actively crawling worms (such as the Phyllodocidae) with a long series of similar segments, each bearing a pair of parapodia with numerous setae which are used in swimming, walking, and burrowing. The next largest group is the class Oligochaeta, numbering more than three thousand species of mainly land and freshwater worms. Earthworms and freshwater oligochaetes have fewer and smaller setae on the segments, never have flaplike parapodia, and usually have head and sense-organs reduced if compared to the polychaete pattern. A third group is the leeches, class Hirudinea, with a more specialized and much more uniform pattern of body involving attachment suckers and a modified mouth and gut. There are about three hundred known species of leeches in the sea, in fresh waters, and on land—the last largely in the tropics. In leeches, setae are entirely absent: fixed points for locomotion being provided by the suckers. In both the Oligochaeta and Hirudinea, the majority of species are hermaphroditic, and some form of copulation results in cross-fertilization, usually internally. As a result, the eggs produced in these two classes can be relatively large, and the developing young upon hatching are not usually ciliated larvae but rather miniature adults. There are two other minor groups of annelid worms, each numbering about fifty species of little ecological importance: the Archiannelida, and the Myzostomaria. The former are a heterogeneous collection, the genera of which may be primitive survivals or may show secondary degeneracy. The Myzostomaria are relatively specialized parasites, almost without internal segmentation as adults, but undoubtedly derived from some polychaete stock.

The class Polychaeta, besides the least specialized forms, encompasses the greatest diversity of form and function among the annelids. By contrast, within the earthworms, or within the leeches, both anatomy and physiology are somewhat stereotyped.

Form and Feeding in Polychaetes

Phyletic discussion of functional morphology in polychaetes is not possible. The structural and functional adaptations associated with certain specific habits and habitats in polychaete worms are often known to be polyphyletic. Each ecological and physiological pattern appears to have been independently evolved in several stocks of bristle-worms. This is revealed by the state of systematics in the group. There are about fifty distinct families of polychaetes, accepted as such by most systematists, but there is no acceptable grouping of these into orders or agreement on any other hierarchy of interrelationships between them. Apart from differing degrees of tagmatization, much of the functional morphology concerns the sensory and feeding structures of the prostomium and peristomium and variation in the parts and proportions of the parapodia. Perhaps the adaptive value of these structural differences is most easily understood in relation to the ecology of the different worms. However, it should be remembered that ecological similarity need not mean close relationship. For example, microphagous polychaetes constructing fixed tubes have apparently been evolved independently in several stocks and thus reflect convergence rather than close relationship.

As already stressed, archetypic polychaetes can be considered as medium-sized worms with many undifferentiated segments, almost certainly detritus-feeding, and possibly living in muddy, offshore bottom deposits. Both *Polygordius* and the least differentiated phyllodocids, although the latter have a more muscular pharynx which can allow macrophagy, can be regarded as independent survivors of such stocks. (It should be noted that the remaining archiannelid genera, other than *Polygordius,* are certainly unrelated and obviously not primitive. A better understanding of polychaete phylogeny would probably merge the Archiannelida into a few families of the Polychaeta.)

Several groups of larger worms—carnivores and macrophagous scavengers—have evolved from these, the principal modification being the development of eversible buccal and proboscis structures along with small cuticular teeth, or stout jaws, or both. Such free-living worms used to be designated as the "Errantia" but involve many distinct and clearly unrelated families including the Syllidae, Eunicidae, Nephthyidae, and Nereidae. Ragworms or clamworms like *Nereis* show the characteristic features and are readily available. There is a high degree of cephalization (see Figures 2·4A and 3·1A), the prostomium bearing four eyes, two tentacles, and two palps, while the peristomium has four pairs of tentacles in addition to the oral apparatus. The pharynx, shown everted in Figure 3·1A, bears stout jaws and many paragnaths or small cuticular teeth. In *Nereis* and similar forms, there are about 80 segments behind the head, all of which are uniform except for the slightly modified first and last. The structure of a regular segment is shown in Figure 3·1B, which, in addition to the internal arrangements of muscles, circular, longitudinal, and oblique, shows the structure of a relatively undifferentiated pair of parapodia. Among the many setae of each parapodium are the stouter acicula which provide a skeletal stiffening, and there are sensory outgrowths called cirri. Most undifferentiated polychaete parapodia are biramous, consisting of a dorsal notopodium and ventral neuropodium (see Figures 3·1B–F), and variations in their proportions allow these paired appendages to be used as swimming paddles, or for creating water currents in tubes, or for crawling or burrowing. In many diverse groups of polychaetes there are also additional lobes, conspicuously blood-filled, which act as gills.

The evolution of a different pattern of eversible proboscis characterizes such forms as lugworms like *Arenicola,* where the "proboscis" so formed is involved both in digging through the mud and in ingestion of suitable organically rich deposits. When lugworms are living in their characterstic U-shaped tubes and feeding normally, rhythmic movements of the everted proboscis are used, not only to shovel into the mouth the richer surface sand which flows down the funnel at the head of the burrow, but also to pump water with suspended material into the gut. The extruded proboscis is also used extensively in the construction of each new burrow, the rhythmic cycle of extension, expansion, and longitudinal contraction effectively pulling the worm through the substrate. *Ophelia* and its allies form another family of similar stout-bodied worms—though some are spindle-shaped—which live in sand and feed, using a similar eversible proboscis.

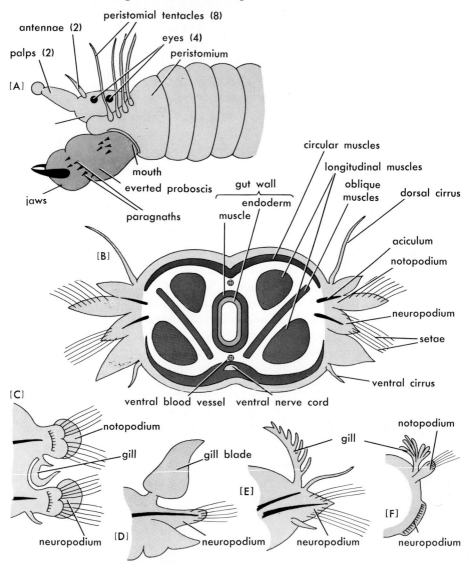

Figure 3·1. Organization of polychaete worms. A: Head structures—prostomium and peristomium—in a nereid polychaete, shown with the proboscis everted. **B:** Generalized nereid segment showing the characteristic arrangements of internal muscles and a pair of typical nereid parapodia. **C, D, E,** and **F:** Stylized diagrams of other parapodia; from a nephthyid **(C)**, from a phyllodocid **(D)**, from an eunicid **(E)**, and from an arenicolid **(F)**.

A vast number of different polychaete groups have developed ciliated tentacles on the head which are used in various methods of particle collection and feeding. The Ampharetidae and Terebellidae have ciliated tentacles on the prostomium which are used to gather materials to build permanent tubes or burrows as well as to collect food. If a terebellid is carefully observed, three methods of food transport are seen to occur. In a terebellid like *Amphitrite,* only one side of each tentacle is strongly ciliated, and this face is usually infolded to form a gutterlike groove in which the cilia beat toward the prostomium and mouth. Fine particles are transported by the cilia; medium-sized particles are maneuvered by the cilia along the groove but their passage back to the mouth is assisted by muscular peristalsis of the gutter (see Figure 3·3B). The largest particles ingested are carried to the mouth by being enfolded in a single tentacle which is then almost entirely retracted.

Figure 3·2. Fan-worms extended and feeding: living specimens of *Sabella pavonia* with, at lower right, a solitary sea-squirt, *Phallusia mamillata*. [Photo © Douglas P. Wilson.]

The polychaetes have also developed true suspension-feeding forms, once again in several families and involving two entirely distinct types of mechanism. As already mentioned, *Chaetopterus* is a suspension feeder using a mucous bag for filtration. The greatly differentiated body occupies a U-shaped burrow with a secreted lining (see Figure 2·4B). Segments 14, 15, and 16 bear the notopodial "fans" which

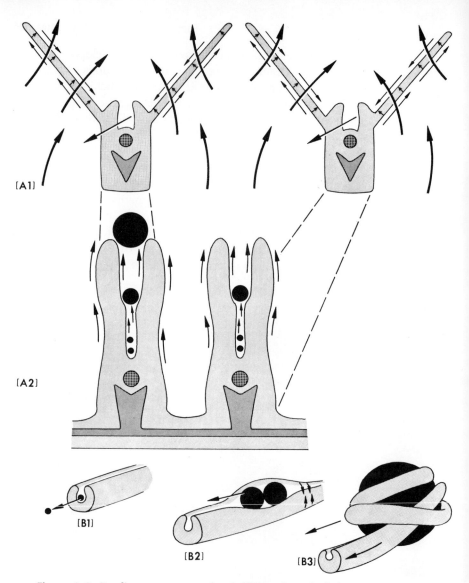

Figure 3·3. Feeding apparatus of sabellid and terebellid worms. A1 and **A2:** Horizontal sections through two adjacent filaments of the feeding fan in a sabellid, distally **(A1)** showing the ciliary tracts on the pinnules of the filaments, and more proximally **(A2)** showing the "size-sorting" apparatus and fused skeletal elements. For further explanation, see text. **B1, B2,** and **B3:** Methods of food collection exhibited by the tentacles in a terebellid like *Amphitrite*, including the ciliary action on small particles in the temporary "gutter" **(B1)**, peristaltic action of the "gutter" on medium particles involving sequential contractions in transverse muscles within the tentacle **(B2)**, and the wrapping action round large particles which is followed by contraction of longitudinal muscles in the tentacle **(B3)**. [Adapted in part from E. A. T. Nicol, in *Trans. Roy. Soc. Edinburgh,* 56:537–598, 1930; R. P. Dales, in *J. Mar. Biol. Ass. U.K.,* 34:55–79, 1955; and unpublished work of Dr. Meredith L. Jones of the Smithsonian Institution.]

create a water current through the tube. Segment 12 manufactures the thimble-shaped bag in which particles are trapped, the bag being ingested every quarter of an hour or so. This filtering mechanism can certainly collect organic particles as small as 1 micron in length, and some workers have claimed that protein molecules of about 40 Ångstroms can be retained.

A totally different type of filter-feeding mechanism is provided by the elaborate tentacles of the fan-worms, the Sabellidae and Serpulidae (see Figures 3·2 and 3·3A). In forms like *Sabella,* the crown of tentacles forms a wide funnel with the mouth at the bottom; cilia on the sides of the pinnules which line each tentacle cause a water current to pass centripetally into the funnel, food particles being trapped, not by a filtering on the outside of the funnel, but by eddies forming in front of, and between, the pinnules. Particles are passed by ciliary tracts toward the base of each tentacle where there are folds forming a sorting mechanism (see Figure 3·3A2). In at least a few species of sabellids, this sorting is threefold: large particles which cannot enter the folds are carried to rejection tracts; medium particles pass along an intermediate groove and are transferred to a ventral sac to become the "graded" material used in tube-building; and lastly, the finest particles pass along the deepest grooves, the cilia of which are continuous with tracts leading directly into the mouth. All fan-worms show some degree of tagmatization. This is perhaps most marked in the serpulids, which secrete permanent calcareous tubes (on rocks, dock pilings, ships' bottoms, and other hard substrata) and have part of the tentacular crown modified as an operculum which can close off the door of the tube when the serpulid is fully retracted.

Excretory and Circulatory Organs

Compared with other wormlike animals—mostly acoelomate or pseudocoelomate in organization—the body of annelids can be both more massive, and more actively employed. Thus the organs of respiration, excretion, and circulation have occasion to be more highly organized than in the other "worms."

The archetypic circulatory arrangement can be best understood with reference to Figure 3·4A. The contractile elements consist of two longitudinal vessels: a dorsal aorta in which the blood is propelled anteriorly and a ventral aorta in which the blood is propelled away from the head. The circulation is a closed one, these two longitudinal vessels being linked by metamerically arranged sets of vessels which join them through capillary systems. Functionally, there are three capillary networks: one on the gut, where nutrients pass into the bloodstream, one in the parapodia where oxygen is taken up, and the third in the muscles where all the usual metabolic exchanges can take place. The afferent blood vessels to the capillaries of the parapodia and muscles leave the ventral vessel in each segment and the flow is toward the dorsal side. In contrast, in each anterior segment the afferent supply to the gut capillaries is from the dorsal vessel, the collecting vessels running to the ventral. However, to complicate archetypic concepts, directions of flow in some "errant" forms have been reported to be exactly contrary to the

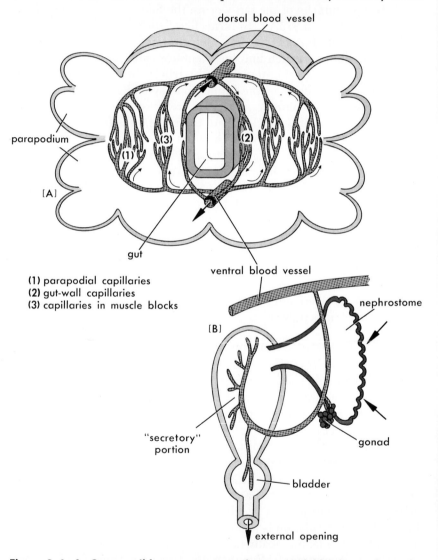

(1) parapodial capillaries
(2) gut-wall capillaries
(3) capillaries in muscle blocks

Figure 3·4. A: One possible arrangement of segmental blood vessels in an archetypic polychaete. For further explanation, see text. **B:** One of the six pairs of excretory organs (mixonephridia) in *Arenicola*. (Compare with Figure 3·5.)

above. In forms showing a higher degree of tagmatization (see pp. 18 and 19), specific groups of segmental circulatory arcs are greatly modified. In some of the more specialized annelids, there is a tendency for the segmentally arranged lateral vessels to become contractile "hearts." At the anterior end of earthworms like *Lumbricus,* there are five pairs of these which pump blood collected from the dorsal aorta through channels with valves to the ventral blood vessel which is, as in all annelids, the principal distributing vessel of the body.

Our present understanding of comparative anatomy, development, and evolution of excretory systems in many-celled animals is largely based on the detailed studies of E. S. Goodrich, which extended over half a century and which were in large part based on the diversity of conditions which are found in various annelids. As Goodrich first elucidated, primitive, coelomate animals, including some surviving annelid types, have two distinct sets of tubular organs connecting the coelom to the exterior. In metameric animals, these are arranged segmentally, archetypically a pair of each in every segment. First, the coelomoducts are mesodermal and grow outward through the body wall from the coelomic cavity. Their archetypic function seems to have been the transport of gametes from the gonads to the exterior. Secondly, the nephridia are developed centripetally from the ectoderm and are primitively blind at their inner or coelomic ends. Each nephridium was archetypically concerned in excretion: as is usual in animals, both in the removal of nitrogenous waste and in osmotic and ionic regulation. The structural distinction will be best understood with reference to Figure 3·5. There are two principal forms of nephridia: protonephridia, where the canals end blindly in either flame-cells or solenocytes, and metanephridia, where the duct system opens to the coelom. Protonephridia are assumed to be the more primitive, and besides being found in such polychaete families as the Phyllodocidae and Nephthyidae, are also the main excretory organs of flatworms and nemertines (see BLI) and are found in *Amphioxus* though not in any other primitive chordates (see p. 183). In one annelid family, the Capitellidae, the primitive metameric distribution of the organs is found, each segment having a separate pair of coelomoducts and metanephridia. In all other annelids and in many other phyla, the two sets of structures are combined in various ways, and, once again, the details of the different types of fusion which can occur were first deciphered by Goodrich. The more important combinations are illustrated in Figure 3·5C, including the protonephromixium as in Phyllodocid worms, involving combination of a protonephridium with the coelomoduct, and two different forms of combination of metanephridia with functional genital ducts— metanephromixia and mixonephridia. Mixonephridia, the last form of combined segmental organs, are found in such worms as *Arenicola,*

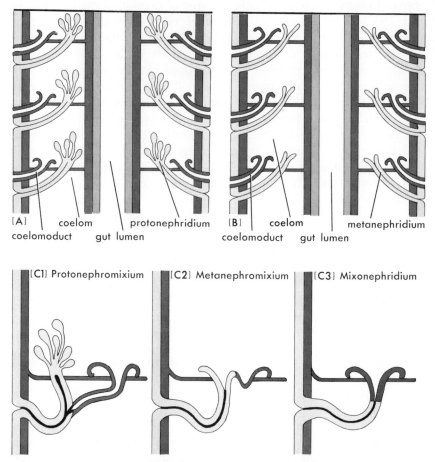

Figure 3·5. Segmental organs (nephridia and coelomoducts) in annelids. A: A presumed archetypic condition with separate pairs of protonephridia and coelomoducts in each segment. **B:** Another archetypic condition, actually found in some capitellid polychaetes, with separate pairs of metanephridia and coelomoducts in each segment. Note that, in both cases, the nephridia are ectodermal and centripetally developed while the coelomoducts are mesodermal and develop outward from the coelomic cavity. **C1, C2,** and **C3:** Three of the more important of the combinations of nephridia with coelomoducts which are found in various polychaetes: a protonephromixium, a metanephromixium, and a mixonephridium. For further explanation, see text.

where they are restricted to the 6 segments of the second tagma of the divided body. Functionally, they are concerned with both the passage of gametes to the exterior and with nitrogenous excretion and ionic and osmotic regulation (see Figure 3·4B). They are obvious organs in a freshly dissected lugworm, each with a frilly funnel as an internal opening, a rich supply of blood vessels, and the gonad tissues in close proximity. All oligochaetes and leeches have metanephridia. In these groups, the coelomoducts with the gonads are restricted to a limited number of segments and are not usually connected with the nephridia. Obviously the functional importance of nephridia is greater in nonmarine annelids. In polychaete groups, where there are estuarine and freshwater forms, these have nephridia that are not only larger and more complex than the corresponding marine species but also have a richer vascularization. Physiological investigation has been most complete in earthworms. In most of them, the fluid discharged by the nephridia is hypotonic to the coelomic fluid (i.e., contains less dissolved salts), and has a higher proportion of such nitrogenous substances as ammonia and urea. At present, there is some controversy among physiologists about the functional processes in annelid nephridia.

Regeneration and Reproduction

The pattern of morphogenesis of segments in annelids, described in the preceding chapter, is associated with a fantastic capacity for repair after injury by regeneration of lost segmental units. In some polychaetes and aquatic oligochaetes, fragmentation is combined with regeneration as a method of asexual reproduction. The development of specialized sexual individuals, each produced asexually, seems also to have arisen from the regenerative capacity.

Restorative (or reparative) regeneration can occur to some extent in all annelid worms, and usually returns the injured worm to the characteristic number of segments of its adult form. The restorative process is easiest to understand in worms where all the segments are relatively uniform. It should be remembered that the actively growing zone in young worms is usually just in front of the most posterior segment. In many forms regeneration can take place headwards or tailwards, as appropriate, the wound surface first sealing to form a blastema, which then proliferates an appropriate number of segments to restore the adult pattern. It is significant that in restorative regeneration of the anterior region, it is the head which is differentiated first, and in many elongate worms this leads to some imperfections in the restorative process. In such forms, if the head plus from 3–5 segments is cut off, then the head plus 3 segments only regenerates; if the head plus from 6–10 segments is cut off, then the head plus 4 segments regenerates; and if the head plus 20–30 segments is cut off, then only a head plus 10 segments is restored. In the elongate syllid *Autolytus*, a cut made between segments 5 and 13 will cause restoration of a head plus 5 segments; amputation between there and segment 42 results in restoration of a head only; and with a more posterior cut, no head can be regenerated. In several worms, there is a specific regeneration no matter how many segments are removed. For example, in *Syllis spongicola* the restorative regeneration is always of the head plus 2 segments no matter where the amputation. Perhaps surprisingly, even worms with a high degree of segmental specialization (tagmatization) are capable of restorative regeneration. Any number of segments up to 14 may be cut off the anterior end of the highly specialized *Chaetopterus* and complete restoration can occur. If 15 anterior segments are cut off, there is no regeneration. Fantastically, any 2 segments from the anterior tagmata will regenerate both anteriorly and posteriorly to form an exact copy of the original highly-specialized worm.

Tube-dwelling fanworms like *Sabella* must naturally be rather subject to loss of the tentacular crown and anterior segments which are exposed from the tube during feeding. Restorative regeneration in *Sabella*, after the head structures and any number of segments are removed, always results in the head (prostomium and peristomium) plus one body segment. Subsequently, the existing segments are modified by metamorphosis and new ones added posteriorly to make up the number. There is considerable evidence that the restoration of the correct segment order in *Sabella* and similar forms involves secretions passing from the regenerated head structures posteriorly.

Competence in restorative regeneration, particularly the capacity to regenerate a lost head, is linked in many aquatic annelids with the development of asexual fragmentation. In the genus *Ctenodrilus*, worms simply periodically fall apart, and each group of 1, 2, or 3 segments then regenerates into a complete worm. In *Autolytus* and several other genera, a similar asexual process is more closely controlled, with epidermal ingrowth forming a macroseptum at the position where the break will occur and a head forming behind each macroseptum before fission takes place. A chain of stolons may thus be formed (see Figure 3·6A) and can result in the budding off of nearly complete and fully grown "adults."

A special form of this budding of individuals is called epitoky. This involves the development by asexual means of a reproductive individual, called an epitoke, which differs from the nonsexual atoke usually, not only in the possession of mature reproductive organs internally, but also in the possession of modified parapodia to allow more efficient swimming. The ecological importance of epitoky is twofold. It can allow synchrony of spawning, and by providing a motile distribution stage, permit the scattered burrowing worms of the sea bottom to come together in swarms for reproduction.

Both in cases where epitokes are budded off, and in cases where the whole adult worms metamorphose to a sexual swarming form, both the structure and behavior of the sexual forms is markedly different from their normal, bottom-dwelling progenitors or predecessors. Structurally, the sexual individuals have reduced guts, enlarged eyes, and parapodia elongated and elaborated as effective swimming paddles. Behaviorally, sexual individuals are usually markedly photopositive being attracted to artificial lights in appropriate seasons. Scientifically, there are many palolo-worms, several genera and species occurring in different parts of the world. The commonest Pacific palolo belongs to the genus *Eunice,* and its spectacular swarms can be predicted with some accuracy since they show response to the lunar cycle. As a result, these or similar worms are netted as a source of human food in Fiji, Samoa, and certain parts of Japan. More minor, but still impressive, swarming of sexual individuals can be observed in many parts of the world. Two examples are well known to students who have visited Woods Hole on Cape Cod. Both regularly appear in July two nights after the occurrence of full moon. In the case of sexual individuals of *Nereis succinea,* the males appear first and swim in rather loose circles. When females appear amongst them, the males form swirling vortices around each female and only then emit sperms. The sperms apparently elicit egg-shedding by the females, fertilization occurs, and both male and female worms survive the sexual aggregation. In *Platynereis megalops,* sexual swarming involves a unique copulatory process, regularly observed by investigators and students at Woods Hole, but so peculiar that a number of writers on the subject have doubted its occurrence. The eggs in this species are unfertilizable after 40 seconds of contact with seawater. As with the previous species, both male and female sexual individuals swim toward night lights held over the surface waters. On contacting a female, the male wraps his body around hers, inserts his anus into her mouth, and goes into a general contraction of the body muscles. Since in both sexes the gut wall is already degenerate, sperms pass from the male's anus directly into the coelom of the female where the eggs are fertilized and spawned via the female anus almost immediately. In this case the sexual individuals do not survive the process.

Apart from the complete metamorphosis of a bottom-dwelling polychaete into its sexual swimming form, there is great variation in the type of asexual division leading to epitoke formation. Thus some epitokes are long, of many segments, others short; some have heads and obvious sense-organs, others have none. There is considerable evidence, however, that some of the processes of sexual metamorphosis after epitoky can only take place after separation from the original parent head. It seems almost certain that neurosecretory cells in the head of the "normal" worm secrete a hormone which has an inhibitory effect on the development of the secondary sexual characters (including the enlarged parapodia and eyes) of the epitoke and may even control the onset of gonad maturation. It seems likely that a similar endocrine control could be involved in some of the processes of restorative regeneration discussed earlier. Neurosecretory cells and their products will be more fully discussed in the Arthropoda, where their physiology has been more thoroughly investigated.

As already noted, the terrestrial annelids—earthworms and leeches—are hermaphrodite with internal fertilization and relatively large eggs. The functional organization is often complicated, but is essentially concerned with the prevention of self-fertilization while accepting sperms from other individuals to fertilize the eggs. In an earthworm like *Lumbricus* (see Figure 3·6B and C), there are essentially three sets of

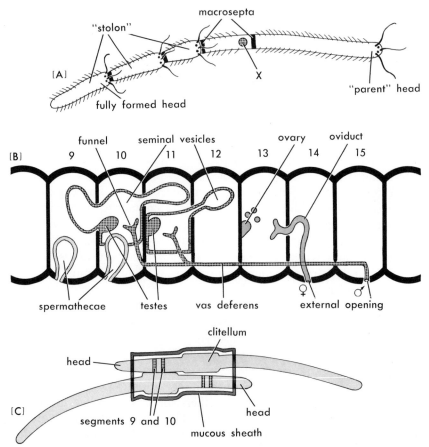

Figure 3·6. Reproduction in annelid worms. A: Stolonic budding in a syllid polychaete, *Autolytus*. After each macroseptum is formed by epidermal ingrowth a new head is differentiated behind it before the asexually budded individual breaks off. **X** marks a region where implantation of cephalic neurosecretory tissue will prevent head differentiation. **B:** Schematic lateral view of the hermaphroditic sex organs in a typical earthworm. **C:** Characteristic arrangement of two earthworms in copulation.

structures: first, two pairs of spermathecae which store the "foreign" sperm received from the partner in copulation; secondly, two pairs of testes, with their storage seminal vesicles and their modified coelomoduct system to the exterior; and thirdly, a pair of ovaries and their modified coelomoducts (see Figure 3·6B). In copulation, two such worms have their ventral surfaces in contact and secrete an enclosing sheet of mucus. Actual contact is between segments 9 and 10, the spermathecal segments, of one worm and the clitellum of the other. Sperm stored in each seminal vesicle pass down the vas deferens, out of the male opening in segment 15, and along the outside of the body to the clitellum where they enter the spermathecae of the other worm. The worms then separate, and the clitellum secretes a cocoon of hardened mucus which slides forward with the wriggling of the worm. As it passes the female opening on segment 14, eggs are extruded into it, and a little later it receives sperm from the spermathecae. Finally it is passed over the head and the ends are closed off; the zygotes develop inside the egg-cocoon to hatch out as miniature adults. The details vary in different terrestrial annelids, but the functional implications are always the same. They involve copulation and cross-fertilization followed by direct development (without larval stages) from relatively large eggs.

Minor Annelids

The minor class, Myzostomaria, is a highly specialized group of parasites, almost certainly derived from polychaetes. Myzostomarians appear as adults like tiny flattened discs with five pairs of reduced parapodia. They are all parasitic on echinoderms, the majority on crinoids. Although they are hermaphroditic, the development of their fertilized eggs leads through a typical free-swimming polychaete larva.

The other minor class in the phylum Annelida is the Archiannelida. As presently constituted, it is certainly polyphyletic. Apart from *Polygordius,* the other archiannelid genera all have muscular buccal bulbs and may severally represent stocks showing simplification (including absence of setae) from several polychaete families. Such genera as *Dinophilus* and *Nerilla* could possibly represent small-scale and degenerate versions of ancient stocks of eunicids and syllids respectively. The common interstitial habitat could well be significant. Even if some of the other genera, including *Polygordius,* genuinely represent surviving primitive stocks, there seems to be a good case to suppress the artificial group, Archiannelida, and incorporate its parts into various subdivisions of the class Polychaeta.

Nonmarine Annelids

All available evidence from comparative anatomy and physiology suggests that the first oligochaetes were evolved from representatives of one of the less-differentiated polychaete stocks which acquired a capacity for osmoregulation and invaded fresh waters. The primitive oligochaetes of fresh waters gave rise to more completely adapted forms with, besides osmoregulatory capacity, internal fertilization, large eggs, and resistant resting stages. The successful freshwater oligochaetes of the present day, including such ubiquitous families as the Naididae and Tubificidae which are ecologically important in lake bottoms, belong to this group. Both the land oligochaetes, or true earthworms, and the freshwater leeches must have been evolved from similar freshwater oligochaete stocks which already possessed the full concert of structures and functions for nonmarine life. For example, evolution of internal fertilization and the production of large eggs preceded the development of earthworms or leeches.

It is worth noting that the evolution of earthworms must have taken place during and after the Cretaceous, in close correspondence (as with the insects) with the evolution of a land flora of flowering plants. Terrestrial soils, as we know them, evolved along with the seasonal and deciduous vegetation of the newly evolved angiosperms, and with the newly evolved land oligochaetes. Angiosperms, earthworms, and humus all have exactly the same geological antiquity.

Locomotion in earthworms involves neurally controlled waves of muscle contraction which pass anteriorly along the worm, adding posteriorly to each bulge where the longitudinal muscles are contracted and the setae protruded. Thus the bulges move slowly back, and the body of the earthworm moves somewhat more rapidly forward. (The process is fully described in BLI, pp. 106–111.) This method of locomotion serves the earthworm both in crawling over the surface (usually at night) and in burrowing through soft soils. Along with the soil itself, earthworms ingest all kinds of organic material but especially leaves and other dead plant tissues. As was extensively investigated by Charles Darwin nearly a century ago, earthworms modify the soil structure by passing soil through their bodies and casting it toward the surface. Essentially, besides improving aeration and drainage of the soil, earthworms move organic material downward deeper into the soil and bring inorganic particles from the subsoil upward toward the surface. Earthworms occur in all parts of the world, under all types of vegetation, and in all climates except the most arid. They form an important part of the terrestrial biomass, apart from their ecological importance in restructuring soil. Under one square meter of pasture grassland, eight hundred larger earthworms, and ten times that number of smaller oligochaetes, can occur. There is probably a greater abundance in the soils of deciduous woodlands.

The most highly specialized annelids are the leeches (class Hirudinea). Although they are found in the sea as well as on land and in fresh water, it is probable that the marine forms, like the terrestrial ones, are evolved from freshwater leech stocks, which in turn were evolved from freshwater oligochaete stocks. They form a very stereotyped group in which the two suckers have replaced setae as the fixed points in locomotion and where the coelomic cavity has been largely obliterated by the growth of fluid-filled cells. The mouth is in the center of the anterior sucker and may be armed with three piercing jaws. It leads into a muscular pumping pharynx and thence to a large storage crop which is functionally related to the irregular meals. Leeches, like earthworms, are hermaphrodites with the organs arranged for cross-fertilization. Most leeches are not true parasites since they remain attached to their hosts only for the short period of feeding. Many leeches are predaceous carnivores rather than blood-suckers and eat such invertebrates as earthworms, slugs, and insect larvae. As is well known, in the blood-sucking forms, the salivary glands secrete an anticoagulant which is injected to prevent clotting of the host's blood during feeding, and is responsible for the continuation of bleeding after the swollen leech has withdrawn.

A few leeches have a capacity for restorative regeneration of the head but there is no asexual reproduction in the group. This is probably connected with the highly stereotyped body plan; except for one genus, they have a body invariably composed of 33 segments, the last 6 of which form the large posterior sucker. The Hirudinea undoubtedly comprise the most specialized, and least diversified, group of annelids.

The possible relationships between other phyla of many-celled animals and the Annelida are discussed in Chapters 7 and 14.

4

Functional Organization
of Arthropoda

ON THE BASIS of mechanical functioning and of inferred phylogeny, the characteristic features of arthropods can be divided into two distinct lists. There are those paralleled in the phylum Annelida (mainly involving aspects of development and metamerism), and those more associated with the possession of a chitinous exoskeleton. The features on the first list would include their being triploblastic; coelomate; with perfect bilateral symmetry; with a tubular gut running from the mouth to the terminal anus; with metameric segmentation as we have already defined it; and with the central nervous system in the protostomous form (see Chaper 9, p. 131), that is as a chain of segmental ganglia on the ventral side with only one pair of ganglia—the supraoesophageal ganglia—lying anterior and dorsal to the gut. The second list would include the many features which are associated with the exoskeleton of chitin and epicuticle: such as the need for periodic molting or ecdysis during development and growth; the jointed limbs with their internal arrangement of antagonistic muscles as flexors and extensors; the restriction, compared to annelids, of suitable permeable surfaces for respiration, excretion, and so on; the contrasting capacity for resistance to water losses, which has made them the most successful group of terrestrial animals; tagmatization (as already defined) occurring to a much greater extent than in the annelids; an apparent lack of ciliated epithelia even in the linings of the internal duct systems; and the body-cavity as a hemocoel involved in the circulation of the blood, rather than as the annelid coelom, which is totally separate from circulation and greatly involved in the mechanics of muscle antagonism.

Thus all the features truly diagnostic of the arthropods are connected with the exoskeleton—even the most unlikely diagnostic characteristic, "hearts with ostia," is related to the passage of blood from the hemocoel into the heart and the lack of a "venous" system, and thus to the decline in the importance of the hydrostatic function of the body-cavity fluids.

By any measure, the phylum is enormously successful, clearly the most successful invertebrate phylum, whose species numbers (see Table 1·1) outnumber those of all the other phyla added together. The arthropod fraction of the total animal biomass is likewise disproportionate: consider, for example, the composition of the faunas of terrestrial soils or of the marine plankton. Physiological aspects of this success are, once again, almost all associated with the possession of an exoskeleton of chitin and epicuticle. As noted in the introductory chapter, the account of arthropod functional organization below and the chapters which follow are principally concerned with the biology of the arthropod class Crustacea. Since the other nine classes will be neglected until Chapter 7, and then dealt with somewhat synoptically, it is worth reviewing their relative importance now.

As outlined in Table 2·1, the forms included in the phylum Arthropoda in the strict sense fall into ten groups, here termed classes, though considered as subphyla by some authors. Of these, four are tiny groups of minor ecological importance: Merostomata, Pauropoda, Symphyla, and Pycnogonida. One relatively large group, the class Trilobita, is represented only by fossils, although it has considerable significance as a likely stem group for many, if not all, the other arthropod classes. The classes Chilopoda and Diplopoda form, along with the Pauropoda and Symphyla, a moderately large-sized collection of forms (about nine thousand five hundred species) usually termed centipedes and millipedes, but of relatively minor ecological importance except in some rather specialized terrestrial environments. This leaves us with three enormous classes: Crustacea (more than twenty-eight thousand species), Arachnida (more than forty-seven thousand species), and Insecta (more than seven hundred thousand species). The crustaceans are a largely marine group, with considerable representation in fresh waters and only a few genera on land. On the other hand, the arachnids and insects are primarily land animals with only a few forms secondarily returning to aquatic life, usually in fresh waters. It is this ecological distribution of the three large classes, and the occurrence of relatively primitive marine forms in several of the subclasses of the class Crustacea that makes a study of the more primitive crustaceans relatively more rewarding in terms of an understanding of the possible functional homologies and the course of physiological evolution in the arthropods as a whole. Their greater physiological diversity contrasts with rather stereotyped patterns of anatomy and physiology in the arachnids and insects. Since we cannot investigate aspects of physiology in the class Trilobita, the closest approximation to archetypic conditions of structure and function is likely to be found in the more primitive marine Crustacea.

Physiology of the Exoskeleton

Structurally, the layers of the arthropod exoskeleton are divided into two zones, and the functional significance is markedly different for each. The thin, outer nonchitinous epicuticle is responsible for the "chemical" properties of the integument. The much more massive chitinous procuticle provides the mechanical properties of this combined armor and skeleton. The procuticle consists of many fused lamina which can be grouped together on the basis of differences in composition or properties. Unfortunately, the nomenclature adopted by workers on insects does not correspond to that used by most workers on the higher Crustacea: the former dividing the chitinous layers into an exocuticle on the outside next the epicuticle and an endocuticle below this, while the latter describe the whole procuticle as the endocuticle, dividing it from outside inward into a pigmented layer, a calcified layer, and an uncalcified layer (see Figures 4·1A and B). Both the epicuticle and the

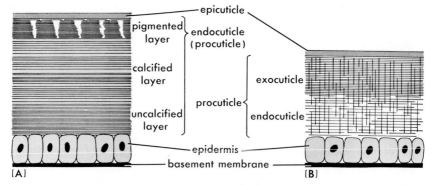

Figure 4·1. Structural patterns in arthropod integuments: A, in decapod crustaceans and **B,** in insects. The functional significance of each of the different layers is discussed in the text.

chitinous procuticle consist, of course, of nonliving layers secreted by the cells of the ectodermal epidermis which lie within and in close contact with them. This epidermis has typically a stout basement membrane to which muscles can be attached and is itself strengthened by intracellular fibrils which run from the basement membrane into the cuticle.

The epicuticle is responsible for the "chemical independence" of the environment which the exoskeleton confers on arthropods, being markedly hydrofuge (or nonwettable), almost impermeable in a physicochemical sense, and providing protection against microorganisms such as bacteria. The details of the ultrastructure responsible for these properties, and the sequence of secretion of units, have been most thoroughly studied in insects, but it is obvious that similar organization of the epicuticle occurs in arachnids and in the higher Crustacea. In insects, the outermost layer of the epicuticle consists of a very thin protective cement layer of lipoprotein and is secreted last (actually after molting), the secretion being through pore canals from the epidermal cells which pass through the layers of the procuticle and the deeper layers of the epicuticle. Just below this, lipids are laid down as a waterproof wax layer secreted immediately before molting. Below this again, and next the outer face of the chitinous procuticle, is a layer of proteins—the cuticulin layer associated with polyphenols.

Below these epicuticular layers lies the great bulk of the exoskeleton, the procuticle, consisting of complex laminae of chitin and proteins. Chemically, chitin is a polysaccharide consisting of acetyl-glucosamine residues which are polymerized into long, unbranched molecules. Enmeshed with the chitin in the laminae of the procuticle is a protein component which has been termed arthropodin. Layers of protein and chitin which are not further hardened remain flexible but nonelastic, and such procuticle is found in all the joints of the arthropod limbs and body, and is present to a varying extent (see Figure 4·2) in the innermost layers of the exoskeleton of higher arthropods. Other layers are hardened into a more massive armor or more rigid skeleton. In the crustaceans, this is accomplished by the deposition of calcium carbonate in the middle layers of the endocuticle. Another method of hardening—used to some extent by all arthropods—has been sclerotization and involves the so-called tanning of the protein component in the procuticular layers. This is accomplished by the cross-linking of the protein chains by orthoquinones, and this process always involves, in addition to the proteins, polyphenols and polyphenol-oxidases. In insects, it is the exocuticular layers and in crustaceans, the pigmented layers of the endocuticle, which are most highly sclerotized.

Of course, the mechanical strength conferred by these hardening processes is not important merely in providing a protective armor. The development of an integument of pieces of hardened procuticle which can resist deformation linked by unsclerotized joints is the basis of the superbly efficient arthropod locomotory machines, all of which are constructed as interacting systems of levers. Just as metameric subdivision of muscles and hydrostatic skeleton can increase the mechanical efficiency of soft-bodied worms over similar forms with unsegmented muscle arrangements (see pp. 11, 20, and BLI), so the development of arthropodan levers makes locomotion even more efficient in terms of energy expended. The nice application of force to the substrate or environmental medium is accomplished by the jointed appendages of arthropods as lever-systems, with less waste of energy and with considerably greater precision, than the forces resulting from the contractions of the muscular body wall of soft-bodied animals. Once again, an analogy with man-made machines can be conceived. As always, the basic structures capable of contraction—the motors of the machine— are muscles: circular and longitudinal in the wormlike animals, and arranged as flexors and extensors in the arthropod. The more sophisticated mechanical design of arthropods, however, allows a greater force to be exerted more precisely against the environment, per unit volume of muscle, than in worms. As a result, arthropod locomotion can be faster, more precisely directed, and much more efficient in terms of energy expenditure.

The properties of the exoskeleton as a whole are responsible for the enormous success of the arthropod body plan. To recapitulate, the epicuticle is responsible for the physiological separation of the arthro-

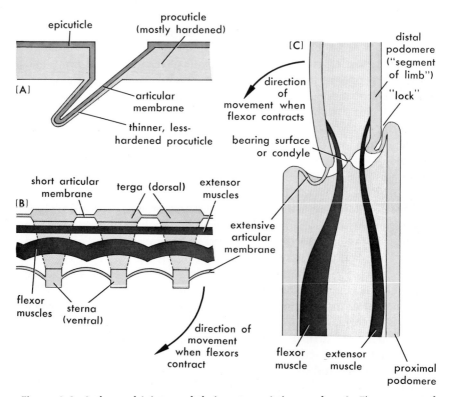

Figure 4·2. Arthropod joints and their antagonistic muscles. A: The structure of an articulation. **B:** Three segments of a schematic half abdomen from a decapod crustacean such as a lobster or crayfish, viewed laterally. **C:** Diagram of a limb joint in a higher crustacean, split open at right angles to the axis of articulation. For further explanation, see text.

pod's tissues from its external environment. The mechanical properties of the procuticle are basic to the precise and efficient movements which characterize the various forms of locomotion in arthropods. Both the physico-chemical and the mechanical properties have obviously contributed to the success of arthropods as land animals. It is worth noting that no other invertebrate group is so successful on land, and further, that the greatest proportion both of terrestrial species and of terrestrial animal biomass is arthropodan.

Joints, Muscles, and Limbs

The characteristic joints of arthropod bodies and limbs depend on their being regions of the procuticle which are thinner, less hardened, and thus lacking the mechanical rigidity of the other parts of the integument. It should be noted that these articular membranes, as they are called, are never intrinsically elastic, although this is suggested in many textbook accounts. In general, the articular membranes are more extensive on one side of the joint than the other (see Figure 4·2). Thus the abdominal segments of the crustacean in Figure 4·2B will be capable of greatest bending ventrally, while those of the joints in Figure 4·2C will bend to the left. The actual "bearing surfaces" are termed the condyles, and the invaginated exoskeleton provides points—apodemes —for the attachment of the muscles. Mechanically, each podomere (segment) of a stenopodous limb (such as the walking legs of insects, spiders, or Crustacea; see below p. 105) is in effect a rigid tube linked by the articular membranes to the next rigid tube. A line through the condyles, where each tube actually touches the next, forms the axis of articulation of the joint, and normally there are muscles arranged across the joint on either side of this axis in antagonism to each other.

The functioning of the complex internal musculature of arthropods, and the mechanics of arthropodan movements, are most readily understood if two simple facts of muscle physiology are fully accepted. First, the muscles in any arthropod exert forces roughly proportional to their size (to volume of contractile fibers). Secondly, arthropodan muscles— like the muscles of any other animals—can exert force only by contracting. (No mechanism which involves muscles in exerting force by "expanding" or "stretching" can possibly exist in animals.) A muscle is normally caused to contract by a stimulus from a nerve or nerves. As soon as the stimulus from the nervous system ceases, the muscle will become limp again, but the unstimulated muscle does not regain its original, precontractional length until it is extended by some other force outside itself. In arthropods, as in most other animal mechanisms, this force is provided by the contraction of another muscle or set of muscles which are said to be antagonistic to the first. The action of antagonistic muscles around the joints of the *internal* skeleton in vertebrate limbs is familiar, and the antagonism of circular and longitudinal muscles around the fluid-filled bag of a hydraulic skeleton in wormlike animals has been discussed elsewhere (see pp. 11 and 141, BLI, pp. 104–111). In arthropods, the antagonists span each joint on opposite sides of the axis of articulation. In general in the great bulk of arthropod systems, the muscles—the flexors—lying on the side with the greatest extent of articular membrane are the more massive. The muscles crossing the other side of the joint—the extensors—are usually slighter. Again in a generalization to which there are some exceptions, it is the flexors in arthropod systems which do most work against the environment, and the extensors are the "recovery" muscles. (Many of the exceptional cases, where extensors are propulsive, occur in land arthropods; see Chapter 7.) In very general terms, therefore, flexors tend to be distributed on the ventral, or posterior, or medial sides of the joints of individual limbs or the body, and are normally the more bulky muscles. (In the analogous mechanical system of the muscles working round the elbow joint of the human arm, the biceps brachii—with which more work is done—is much more massive than the corresponding antagonist which straightens the arm, the triceps brachii.) In many higher crustaceans, like the crayfish, the abdominal segments are arranged as in Figure 4·2B. These crustaceans often show an escape reaction involving flexure of the "tail," which drives them backward away from any danger. This flexure is carried out by the more ventral set of longitudinal muscles in the abdomen (see Figure 4·2B) which are more massive, while straightening of the abdomen is accomplished by the slighter dorsal longitudinal musculature. Another characteristic feature of the functioning of arthropod muscle systems is illustrated in Figure 4·2C, where on contraction of the extensor muscles, the rigid elements on that side of the skeleton, being separated by much less extensive articular membranes, are so arranged that they lock together rigidly on contraction of their muscle. In a minority of arthropodan joint systems, there are no extensor muscles in the strict sense, and the flexors are stretched by other means. Elastic structures counter them in a few cases, while in others (including the mouth parts of butterflies, the cirri in barnacles, and the limbs of certain peculiar arachnids), extension is accomplished by hydraulic means. In these peculiar cases, bending of the structures is accomplished by the flexor muscles but these are stretched again and the structures are extended by blood being pumped into them. A few zoological textbooks claim this extension mechanism to be universally important in the mechanics of arthropod movements, but this is not so. The usual antagonists of flexor muscles are extensor muscles. In the claws, or chelae, of crabs and lobsters, the closing muscles are flexors and the opening ones extensors.

A simple mechanical experiment with a healthy crab or medium-sized lobster will verify the truth of the statement above regarding the greater force exerted by the flexors. It is relatively easy for human fingers to hold such a claw closed, that is, to oppose the force exerted by the extensor muscles, but much more difficult to hold the same claw open, that is, to oppose the flexors.

Another aspect of the physiology of arthropod movement seems peculiar to us as vertebrates. The nerve-muscle system seen in crustacean limb muscles functions in an entirely different fashion from vertebrate somatic muscles. The extent and nature of contractions in crustacean limb muscles is variable, but the variation results from multiple innervation. Double, triple, and quadruple motor innervation has been demonstrated in different crustacean muscles, and van Harreveld and Wiersma demonstrated some years ago that a quintuple innervation of all fibers occurred in certain muscles in higher Crustacea. Experimentally, four of the five nerve fibers were found to be motor axons, each of which elicits a contraction with different characteristics. The fifth fiber, when stimulated simultaneously with any of the other four motor fibers, causes inhibition of the contraction. The muscles in which this quintuple innervation was demonstrated occur in decapods and are the flexor muscles of the carpopodite; that is, they are more distal muscles in the walking legs which are important for the posture of the animal and the support of its body weight.

The limbs of primitive Crustacea—and possibly of all primitive arthropods—are constructed on a biramous pattern. There is a basal protopodite bearing the two rami, the exopodite and endopodite. Primitively, such limbs are generalized in function, being involved in locomotion by swimming or walking, in feeding both in the creation of a water current and in sieving, in respiration (particularly if they involve flat laminae), and as sensory organs bearing receptors. In more advanced arthropods, the limbs are more specialized for one or two of the four basic functions. In crustaceans there are two main lines of modification of the limbs: the phyllopodium and the stenopodium (see Figures 5·1B and C). The names of the individual podomeres, or joints of the endopodite in the stenopodous walking limb, need only be remembered by arthropod systematists, but there are a few terms used in limb anatomy which are concise and valuable in any discussion of functioning systems in Crustacea. Besides the main terms protopodite, endopodite, and exopodite, it will be advantageous to remember that additional lobes developed on the inside of the limb are endites, and those on the outside exites. Endites on the protopodite form the gnathobase, usually used in food collection or trituration, while exites on the protopodite (proximal to the exopodite) are epipodites, which, in many Crustacea, are flattened lobes used as gills.

Two points of mechanical functioning in these two basic limb types are worth noting here. In the typical stenopodium, the axes of articulation vary in direction in successive joints. Thus, although each pair of podomeres can be moved in relation to each other only in a plane at right angles to the axis of articulation, the limb as a whole is capable of nearly universal movements. In some other systems of joints, for example in the sensory antennae of higher Crustacea, the condyles are not present and a specific axis of articulation is not defined. In these, movement is possible in any plane (in theory at least). In phyllopodous limbs, the principal musculature works around the joints of the protopodite, and to a much less extent, around more distal joints in the limb. In many primitive Crustacea, phyllopodous limbs are used both for swimming and creating a feeding current. In these limbs, the joints between the more distal lobes—such as endites and the exopodite—and the central parts of the limb are so arranged that they flex toward the animal until they lie at right angles to the plane of the central parts of the limb, but are arranged with extensor locks (as discussed above) which prevent them from being flexed anteriorly at all (see Figure 5·2B). Thus, purely passive movements involving these joints allow the limb, moved by the contractions of muscles across the joints of its protopodite, to execute an effective backward stroke and a recovery stroke, and achieve effective forward locomotion. On the recovery stroke, the limb is moving anteriorly but all its peripheral flaps are flexed back in a manner analogous to the "feathered" blade of an oar in rowing. While this arrangement of jointing is essential for the locomotion of the animals concerned, it is also of fundamental significance in their feeding mechanisms. This will be discussed later (see pp. 60 to 62).

Molt Cycles

Growth, in animals other than arthropods, can involve a gradual increase in size, accompanied if necessary by gradual changes in shape, until the size and form of the fully grown adult is reached. Possession of an exoskeleton prevents this in arthropods. Once an arthropod exoskeleton has been hardened, by sclerotization and calcification, no change in its external linear dimensions can occur. Growth in arthropods must thus proceed through a series of molts. Each must involve the secretion of a new cuticle and an ecdysis, or shedding, of the old exoskeleton. The actual molting, or ecdysis, takes place *after* secretion but *before* hardening of the new, and larger, cuticle. During the short period of hardening the new exoskeleton, the arthropod shows a rapid increase in bulk, which usually involves uptake of water or air into internal spaces. Then, subsequently, the arthropod grows new tissues to fill the new armor. Thus arthropod growth appears to take place in spurts (see Figure 4·3). It is worth emphasizing the paradox involved in this. It is that while the apparent increase in size in arthropods occurs at the molts (the near-vertical curves of Figure 4·3), almost all of actual tissue growth, or increase in individual biomass of the arthropod, takes place during the intermolt periods when no change in size can be detected externally (the horizontal "steps" of the figure).

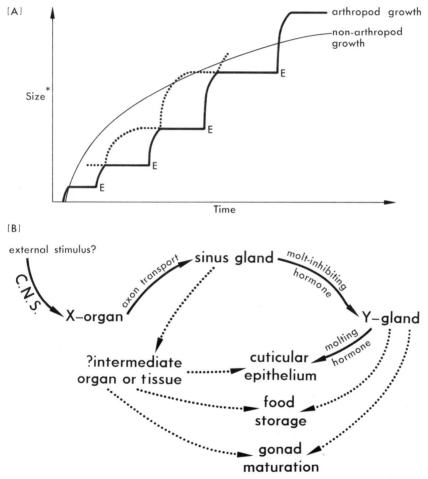

Figure 4·3. A: The characteristic pattern of arthropod growth (size being measured by any linear dimension on the exoskeleton). Rapid size increase follows each ecdysis or molt (E), and results in the growth steps which contrast with the smooth curve for growth in other animals. Paradoxically, almost all real tissue growth or increase in individual biomass (dotted line) takes place during the intermolt period when no change of size can be detected externally. **B:** An outline of the interactions of the neuroendocrine system controlling the molt cycle in higher crustaceans. For further explanation, see text.

Many arthropods have changes of form at several ecdyses, which are termed metamorphoses. In others, including the crayfish, growth from juvenile to adult is accompanied by gradual shifts of proportions and there are no drastic changes at any one molt.

It is worth noting that all surfaces covered with cuticle are renewed at each molt. There are many invaginations of the ectodermal epithelium, and with each is associated a cuticular lining, structurally and physiologically homologous with the exoskeleton. Such invaginations include the fore- and hind-gut, the various apodemes for muscle attachments, and the tracheal tubes of air-breathing arthropods. The old cast cuticle of an arthropod, with all these attached internal processes, is convincing circumstantial evidence that the new cuticle was soft and pliable during the ecdysis.

The physiology of the molt cycle has received most extensive and detailed study in certain insects and in higher Crustacea like crabs. For the latter, an accepted modern classification of the stages was developed as a result of the detailed studies of P. Drach (see Figure 4·4). In summary, there are four functionally different stages covered in the detailed cycle from C4 to E to A1 and back to C4. In proecdysis, or premolt, calcium is removed from the exoskeleton and the calcium content of the blood rises as the new soft cuticle is laid down below the old one. After ecdysis, or the actual shedding of the old cuticle in the molt (Figure 4·5), the crab swells by uptake of water. This is followed by metecdysis, or the postmolt period, during which the exoskeleton is being hardened and calcified. This is followed by the intermolt period when the animal is in normal condition with a hard exoskeleton. This stage (C4) may be long or short, depending in part on the rate of tissue growth and accumulation of organic reserves, this rate in turn depending on the rate of nutrition. In some higher Crustacea there is one intermolt (C4) which is terminal, and the adult animal in these cases can no longer grow. The Drach stages provide a practical classification with considerable detail, into which individual crabs can be fitted on morphological characters and condition of the exoskeleton without previous knowledge of each individual's molt history. Perhaps of greater significance is that Drach's detailed studies have emphasized the biological point: that molting is not a restricted act interrupting "normal" life of the crab, but that all aspects of the normal physiology of the crab are continually changing along with the stages of the molt cycle. For example, as shown in Figure 4·4, the internal water content is varied through the cycle by active processes of uptake and excretion. Similarly, the concentrations of organic food reserves and of mineral salts in the hepatopancreas, and elsewhere in the animal, vary in a pattern corresponding to the Drach stages. Since Drach's work, it has been found that his stages are accurate in detail for all the Brachyura, and also for most other higher crustaceans with thick exoskeletons. Detailed application to smaller Crustacea is difficult but, in more primitive forms, it seems as though the Drach D stages are more extended in time and the C stages shortened.

In all arthropods, there are behavioral changes associated with the molt cycle. Most go into hiding during the period when they are unprotected by the usual exoskeleton (for example, in higher Crustacea,

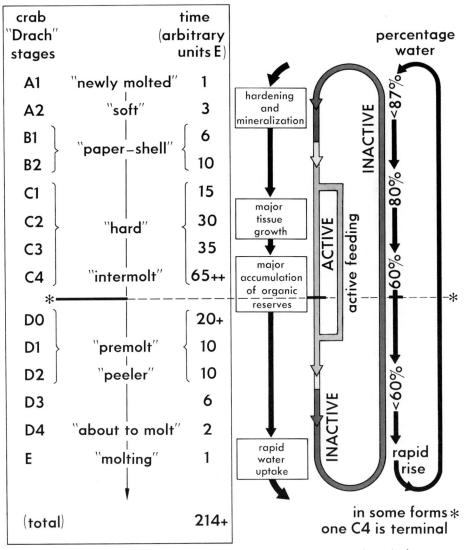

Figure 4·4. Functional aspects of the Drach stages of the molt cycle in higher crustaceans like crabs. For further explanation, see text.

during E, A, and B stages of Drach). In many arthropods, this hiding behavior involves reversal of normal reflex patterns (for example, in light gradients). The molt cycle even affects population dynamics in arthropods. In some forms, a survivorship curve for a population might resemble a mirror image of the arthropod growth curve of Figure 4·4, with a series of steps corresponding to the increased chance of death at each molt, in contrast to the more usual smooth concave curve of survivorship in most nonarthropod invertebrates.

It has long been known that the molt cycle of arthropods is under hormonal control. Once again, the details are best worked out for insects and the higher crustaceans.

Figure 4·5. The process of ecdysis or molting of the old exoskeleton in a swimming crab, *Portunus depurator*. The old integument is above and the newly molted "soft-shell" crab is emerging below. Size increase and hardening will follow. [Photo © Douglas P. Wilson.]

Neuro-endocrine Controls

Our present detailed knowledge of neurosecretory and other endocrine controls of molting and reproductive cycles in arthropods stems from an original discovery made early in this century. It was that in some crustaceans, extirpation of both eye stalks led to precocious growth and, in many cases, to more rapid and repeated molting. Detailed experimental work, particularly during the last twenty-five years, has revealed that three groups of structures are principally involved in higher Crustacea. A hormonal secretion from the Y-glands is responsible for the initiation of proecdysis and thence of the molt. The Y-glands lie either in the antennary or the second maxillary segments of the head. In those

Crustacea with a terminal C4 stage in the molt cycle, the Y-glands degenerate when the mature size is reached. Within each eye stalk is the X-organ which is neurosecretory, and is molt-inhibiting and also inhibitive of the development of the ovaries to functional egg production. The third structure, the sinus gland, also lies in the eye stalk, and is misnamed since it is itself nonsecretory. It involves the endings of axons from nerve cells in the X-organ and elsewhere, receives a secretion from them, and releases this into the bloodstream where it inhibits the activity of the Y-gland. The sinus gland is thus best regarded as an endocrine reservoir.

An outline of these interactions, as they occur in higher crustaceans, is given in Figure 4·3B. At least two secretions are involved. One is the molting hormone produced by the Y-gland, and the other is the neurosecretory material stored in the sinus gland but produced in the cells of the X-organ, which is a molt-inhibiting hormone. There are other organs and secretions concerned in the control of metabolism, of pigment distribution, and of the reproductive cycle in crustaceans. These include the pericardial organ (another neurohemal organ), an androgenic gland formed from cells in the distal wall of the male duct which regulates both spermatogenesis and the onset of male secondary sexual characters in higher Crustacea, and the ovary which secretes hormones directly regulating female secondary sexual characters.

In view of the importance of the neurosecretory material produced by the X-organs, it is worth outlining some aspects of the definition and demonstration of neurosecretory activity. Although neurosecretory cells form part of the central nervous system and are similar in structure to motor neurons, they do not have axons which innervate muscles or any other effector organs in the usual fashion. Their axons mostly terminate in close proximity to a blood vessel or sinus. The cell body often, and the axon at times, contain characteristic stainable droplets or granules which are not present in quantity in other nerve cells. Extraction and injection experiments demonstrate that they contain high concentrations of substances which if circulating in the blood act as hormones, that is, substances of high biological activity on certain other tissues. Thus there are three characteristics involved in their detection: their anatomical position and connections, their histo-chemical properties, and their biological activity if extracted.

It is worth remembering that all nerve cells in metazoan animals are secretory—at their synapses. Therefore, it is often suggested that neurosecretion could have been evolved by a modification of motor neurons. A few investigators—notably R. B. Clark—have pointed out that the evolution of control mechanisms in animals could have taken the opposite course, that is, secretory control in animals could have preceded nervous control.

Brief mention must be made of the control of molting and growth in insects, regarding which some of the most sophisticated experimental studies of hormonal coordination in invertebrates have been carried out. The investigations of C. M. Williams and his associates in the United States, and of V. B. Wigglesworth, J. Harker, and others in Britain are of particular significance. The median region of the brain—the *pars intercerebralis*—forms a region of neurosecretion, but it is probable that there are several different brain hormones with diverse effects secreted here. One of these activates a pair of glands called the prothoracic glands, which in turn secrete a steroid which has been isolated and named ecdysone. This hormone initiates, and in part controls, the concert of functional changes that leads to molting. Another important organ is an epithelial endocrine gland termed the *corpus allatum,* which secretes a hormone termed juvenile hormone. In crude terms, this hormone appears to switch on and maintain the activity of the genes that determine larval form in the insect. In normal circumstances, the juvenile hormone is absent during the last larval molt, and the genes determining adult form can exert their effect. This is an oversimplification: there is obviously a rather complex interaction between the neurohormones produced by the brain and the occurrence of juvenile hormone, in determining the exact degree of change which takes place at each larval molt. Other aspects of function in insects are undoubtedly controlled by hormonal secretions, but it is clear that there is no pattern of endocrine control which is universal for all arthropods, although the use of ecdysone may prove common to all. Ecdysone obtained from insects has not yet been shown to initiate molt in physiologically suitable crustaceans, but extracts of the Y-organ of crustaceans have induced molting in suitable insect preparations, and small amounts of crystalline ecdysone have now been extracted from enormous quantities of shrimp. A difference is that the secondary sexual characters of insects seem to be directly determined by the sex chromosome mechanism and are not, as in crustaceans, affected by hormones derived from other tissues.

Other Functional Systems

Before passing to a survey of functional morphology in the varied extant groups of the class Crustacea, it is worth outlining certain general features in the structure and function of the internal organs in the group. This is possible because there is considerable constancy in internal organization in crustaceans, and also in insects and arachnids. Unlike conditions in such animals as molluscs and echinoderms, where adaptive specialization of external structures usually involves considerable modification of internal organ systems as well, the fantastic variety of crustacean body forms and limb patterns is based on a rather stereo-typed organization of the internal organ systems. The ecology and mode of feeding of any group of crustacean species are reflected in their limb patterns and body tagmatization, but not to any extent in their circulatory and excretory systems, as *is* the case in the Mollusca.

The central nervous system does show some variation within crustaceans. Its most primitive organization is as a ladderlike cord of ganglia, metamerically arranged. There is a greater concentration of ganglia in the more advanced crustaceans. For example, in primitive forms the supraoesophageal complex is made up of the antennular ganglia plus the optic lobes, while in the Malacostraca this is joined by the antennary ganglia. From this complex, the circumoesophageal commissures run to the suboesophageal ganglia, made up of the an-tennary units alone in most primitive Crustacea, but involving several segments in most of the more advanced groups. In crabs there is a posterior fusion of a large number of the thoracic ganglia. A variety of sense-organs occur in Crustacea, and some will be described in more detail later. They include compound eyes, simple ocelli, chemorecep-tors (some on the distal parts of walking limbs), tactile and vibration-detecting mechanoreceptors, statocysts for gravity detection and response to turning movements, proprioceptors, and stretch receptors relaying both static information about limb posture and registering limb movements, and, in terrestrial forms, sensory receptors capable of detecting humidity changes.

In many crustaceans, the alimentary canal is a simple tube with a limited extent of endoderm restricted to the midgut and an ectodermal, and thus cuticularized, stomodaeum and proctodaeum. In the higher Crustacea we find the shortest midgut and the stomodaeum with vari-ous chitinous modifications including filters and triturating mills. The midgut may have diverticula, in some cases forming a hepatopancreas as the main site both of enzyme secretion and of absorption. The diges-tion is entirely extracellular, the foregut being concerned with mechani-cal processing, the midgut with chemical processing, and the hindgut with the formation of faecal pellets, sometimes formed sheathed in cuticular material.

The circulatory system varies, though functionally it always involves return of blood to the heart through hemocoelic spaces. The heart varies in shape but always has ostia and receives blood through them. It may, or may not, pump the blood out through a system of arteries, but there are never return vessels (veins) but only a series of sinuses. The rate of heartbeat in crustacean hearts varies greatly with environ-mental factors. In many forms, the rate of beat is doubled, and in some, tripled, by a rise in temperature of $10°C$, and some crustacean hearts, in translucent forms, respond to changes in light intensity. In most, the heart is neurogenic, that is, the beat originates in nerve cells on the wall of the heart. In marked contrast to conditions in vertebrates, where they are antagonistic secretions, both acetylcholine and adrenaline, on perfusion through crustacean heart preparations, cause increased rates of beating.

There are usually only one pair of excretory organs, although there is considerable evidence that these are derived from a series of segmental coelomoducts (without nephridia). Adult Crustacea usually have a pair of maxillary or antennal excretory organs, but a few larvae have both, and some others change from one to the other in the course of the life cycle. Since the excretory organs lie in the hemocoel, their duct system is always blind at the coelomic end. In most nonmarine crustaceans, the segmental excretory organs are larger and more elaborate, and this usually involves the addition of a long, intermediate convoluted tubule between the blind end-sac (with its attached labyrinth) and the storage bladder (with its short duct to the exterior). Such addition probably corresponds to the evolution of the kidney tubules in nonmarine chordates. These crustacean segmental organs are clearly involved in osmotic and ionic regulation, and functional aspects have been best studied in freshwater crayfish, where urine hypotonic to the blood is produced. Samples taken from different parts of the antennal gland in *Astacus* have clearly demonstrated that the fluids in the end-sac and labyrinth are iso-osmotic with the blood and become hypotonic during their passage through the intermediate tubule but before reaching the bladder. As is the case with annelid segmental nephridia, the precise mechanisms involved in this process are still a matter of controversy among physiologists. It is possible that the functioning is closely similar to that in the vertebrate kidney tubule, with filtration in the crustacean end-sac and labyrinth being followed by reabsorption of ions in the intermediate tubule. However, the observed changes in ionic concentrations along the length of the crustacean excretory organ could result from secretion of water into the intermediate tubule, although this is less likely.

Crustacean Diversity I

THE DIAGNOSIS of the class Crustacea need only read, "arthropods with two pairs of antennae." Crustaceans have three somites in front of the mouth, the second and third of which each bear antennae, and this is true of no other arthropod group. The seeming triviality of this diagnostic feature again emphasizes the stereotyped structural patterns which characterize the major arthropod classes, in spite of their enormous species diversity. The probable relationships between the arthropod groups will be discussed later (see pp. 122–123). Meanwhile, although some workers suggest that the crustaceans are polyphyletic, recent discoveries make this extremely unlikely. The available evidence suggests that all crustacean groups derive from a common ancestral type, and that that type bore certain close structural resemblances to the extinct arthropod stock, the trilobites.

As already noted, there is a tendency for increased tagmatization in the more advanced forms of Crustacea. As well as this specialization into series of different types with modification of limbs, there is also a tendency for a reduction in the total number of segments. There are up to 40 segments in some lower Crustacea, while the higher Crustacea have a consistent pattern of 19 fully developed segments plus one embryonic one, making a total of 20 segments. The segments of the primary tagmata in a primitive crustacean such as *Triops* run: head—1 plus 5; thorax—11; abdomen—22. For a higher crustacean, a typical decapod (such as a lobster or a crayfish), they run: head—1 plus 8; thorax—5; abdomen—6. This indicates both the reduction of total numbers and the process of increased cephalization.

The appendages borne on these series of segments consist primitively of one biramous pair in each segment, based on the protopodite plus exopodite plus endopodite pattern (see Figure 5·1A). However, there are also three single appendages which can be developed: the labrum or upper lip, the labium or lower lip, and the telson, or tail plate, which may be forked or entire and may itself bear claws or jointed rami. Starting at the anterior end, the typical biramous appendages include the anten-

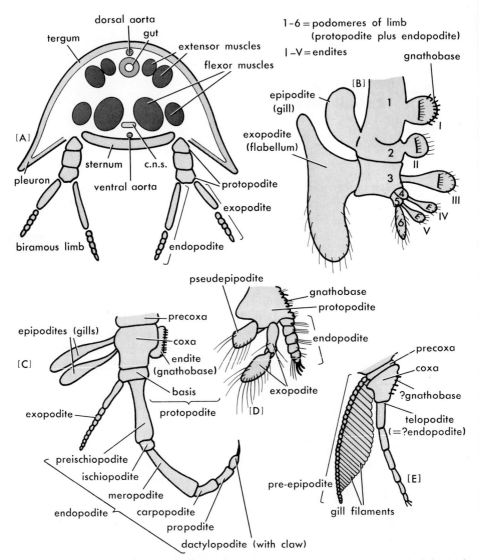

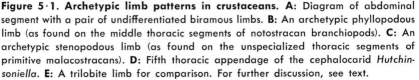

Figure 5·1. Archetypic limb patterns in crustaceans. A: Diagram of abdominal segment with a pair of undifferentiated biramous limbs. **B:** An archetypic phyllopodous limb (as found on the middle thoracic segments of notostracan branchiopods). **C:** An archetypic stenopodous limb (as found on the unspecialized thoracic segments of primitive malacostracans). **D:** Fifth thoracic appendage of the cephalocarid *Hutchinsoniella*. **E:** A trilobite limb for comparison. For further discussion, see text.

nules, or first antennae, which are clearly preoral, usually bear sense-organs, but may be used also in locomotion or in copulation. The antennae or second antennae are in front of the mouth but have nerve connections post-orally; they are usually more clearly biramous than the antennules, but are also primarily sensory in function. The mandibles lack exopodites and each has the precoxa (part of the protopodite) developed as the chewing tooth, while the endopodite forms a tiny mandibular palp. These first three pairs of biramous appendages form the functional locomotory limbs of the nauplius larvae (see p. 14). Behind the mouth are the first and second maxillae, both usually phyllopodous in form. The maxillary segments correspond to the end of the true head, though three further segments and pairs of appendages are added in decapod Crustacea like lobsters and crayfish. The trunk appendages on the thorax are generally called pereiopods, and those on the abdomen pleopods. The first to third pereiopods in the decapods and the first pereiopods in several other groups of crustaceans have become accessory mouth parts as maxillipeds. In *Triops,* there are eleven pairs of pereiopods and twenty-two pairs of pleopods. In the higher forms, there are five pairs of undifferentiated pereiopods and six pairs of pleopods. Particularly in these latter types, the trunk limbs can show great variation in form, being specialized either for feeding, for respiration, for walking or swimming, or for genital and other purposes.

An outline classification of the class Crustacea is presented in Table 5·1. Most textbooks divide the class Crustacea into six subclasses. Recent discovery of living representatives of two primitive genera has required the erection of two more groups: subclass Mystacocarida to include *Derocheilocaris,* and subclass Cephalocarida for *Hutchinsoniella.* Some authorities have placed the latter as the most primitive order within the subclass Branchiopoda. A few authors consider that the Mystacocarida and Copepoda should be combined in one subclass. An old collective term for all the subclasses except the Malacostraca (which constitute the higher Crustacea) was the Entomostraca. This collective term has no phyletic justification whatsoever. Within the Malacostraca, recent reinvestigation of the primitive forms dwelling in caves and hot springs has added a fifth major "division": the Thermosbaenida, and an addition order within the division Peracarida: the Spelaeogriphacea. In this book, reference to the "lower Crustacea" usually involves the Branchiopoda, and the two "new subclasses," Cephalocarida and Mystacocarida. Similarly, reference to the "higher Crustacea" involves the Malacostraca only. Within this last enormous subclass, one order—the Decapoda—encompasses most of the familiar crustaceans: crabs, lobsters, crayfish, shrimps, and prawns. Of the more than twenty-eight thousand described species of crustaceans, over two-thirds fall in the Eumalacostraca, and all but about two hundred of these species in the divisions Peracarida and Eucarida.

TABLE 5·1

Outline Classification of Crustacea

Subclass 1 CEPHALOCARIDA

Subclass 2 BRANCHIOPODA
- Order LIPOSTRACA (fossil only)
- Order ANOSTRACA
- Order NOTOSTRACA
- Order CONCHOSTRACA
- Order CLADOCERA

Subclass 3 OSTRACODA

Subclass 4 MYSTACOCARIDA

Subclass 5 COPEPODA
- Order CALANOIDA
- Order HARPACTICOIDA
- Order CYCLOPOIDA
- (and at least four orders of parasites)

Subclass 6 BRANCHIURA

Subclass 7 CIRRIPEDIA
- Order THORACICA
- Order ACROTHORACICA
- Order ASCOTHORACICA
- Order APODA
- Order RHIZOCEPHALA

Subclass 8 MALACOSTRACA
- Series I LEPTOSTRACA . . . Order NEBALIACEA
- Series II EUMALACOSTRACA (includes five "divisions" of many orders)

Division 1 SYNCARIDA
Division 2 THERMOSBAENIDA
Division 3 PERACARIDA
- Order MYSIDACEA
- Order SPELAEOGRIPHACEA
- Order CUMACEA
- Order TANAIDACEA
- Order ISOPODA
- Order AMPHIPODA

Division 4 EUCARIDA
- Order EUPHAUSIACEA
- Order DECAPODA
  - Suborder MACRURA-NATANTIA
  - Suborder MACRURA-REPTANTIA
  - Suborder ANOMURA
  - Suborder BRACHYURA

Division 5 HOPLOCARIDA (or STOMATOPODA)

In some ways, it is relatively easy to construct a working archetype for the class Crustacea—one with an integrated concert of organs and functions. Apart from involving the features archetypic to all arthropods, mainly those associated with the exoskeleton and its functioning, it would show, as already suggested, relatively little tagmatization or specialization of appendages. It would have a long series of similar segments, each bearing a pair of biramous appendages, probably phyllopodous, all appendages being involved in many functions. It would necessarily be aquatic, with the limbs serving for respiration and the filtering of food particles as well as creating a water current for feeding and locomotion. On the other hand, the construction of hypothetical types is complicated by the actual survival, in certain living crustaceans, of structures and functions as primitive and unspecialized as any in our hypothetical archetype. Such primitive structural features are displayed by the recently discovered cephalocarids. Further, the functioning of a series of unspecialized, multipurpose limbs can readily be investigated in detail in several anostracan genera, which include readily available animals of moderate size, more suitable for experimental work than the minute cephalocarids.

Probably the most primitive living crustaceans are the three species included in the recently erected subclass Cephalocarida. The first of these was discovered and described by H. L. Sanders in 1955 as *Hutchinsoniella macracantha,* the earliest specimens being obtained from an offshore mud-sand bottom in Long Island Sound. The results of a detailed investigation of microanatomy and life-cycle were published in 1957, but the significance of the many unspecialized features was never in doubt. These features include the possession of eight pairs of segmental organs in the thoracic segments (unlike the limitation to one or two pairs in all other Crustacea), the series of nearly uniform paired phyllopodous appendages which include the second maxillae as well as the eight pairs of thoracic limbs (Figure 2·3C), the occurrence of gnathobases on all nine pairs, and the essentially triramous form of all these appendages (Figure 5·1D). Many authorities agree on the general similarity of these limbs to those of trilobites (see Figure 5·1E). Among the more specialized features of *Hutchinsoniella* are the hermaphrodite condition (with separate gonads), and the organization of the 10 abdominal segments which are somewhat cylindrical with pleural spines but no appendages. These features are connected with the nature of the habitat and the small size (about 3 millimeters long).

Obviously, the cephalocarids are closely related to the much more extensive subclass, Branchiopoda, wherein are classified some of the other more primitive crustacean forms as well as some highly successful specializations built on the same ground plan. The evolution of this group has been based on the development of the phyllopodous multi-purpose limbs, and the four living orders represent radiating lines of increasing specialization of the feeding limbs and reduction of their locomotory and respiratory functions. Throughout the four groups there is a tendency for increasing extension of a carapace enclosing the anterior segments. The first order of this subclass shown in Table 5·1 is the Lipostraca, represented only by fossils. The genus *Lepidocaris* includes forms, remarkably well-preserved from the Rhynie Chert in Scotland, of Devonian age, which are similar in size and general organization to *Hutchinsoniella*. However, the thoracic limbs are in two sets, presumably having had differentiation of functions, and there is a pair of very large second antennae which were probably used for swimming as in the modern Cladocera. There is no carapace, and while in some other ways more primitive than *Hutchinsoniella*, *Lepidocaris* also shows some specializations.

The order Anostraca encompasses the least specialized living branchiopods, without any carapace and with the trunk limbs all similar and all used in both feeding and swimming. Genera include: *Artemia*, living in concentrated brine; *Branchipus*, living in brackish water; and *Chirocephalus* and other genera living in fresh waters and known as "fairy shrimps." Most of the species reach lengths of 2 centimeters, while some freshwater forms reach lengths of 10 centimeters as adults.

Archetypic Feeding

Any student who hopes to understand archetypic organization, and particularly feeding processes, in crustaceans should spend some time making detailed observations of the limb movements and the results, in a form like *Artemia* or *Chirocephalus*. Figure 5·2A shows the general body form and the crude movements of water, and of collected food particles, produced by the appendages. The paired limbs are similar (see Figure 5·2B) each with seven endites bearing the filtering setae and with both exites and endites arranged to hinge backward as valve flaps. Each shrimp swims on its back, the same limb movements serving to propel it forward and to collect particulate food. They are true filter-feeders, collecting and ingesting all suitably sized organisms (bacteria, diatoms, flagellates, etc.) suspended in the water and also fine organic detritus. No specialized movements of individual limbs are involved, the backstroke of the limbs being the effective propulsive stroke, while the sequence of events during the recovery, or forward stroke, allows for the filtration of the suspended food particles from the water. This will be best understood with reference to Figures 5·2C1, C2, C3, and D, and depends on the metachronal rhythm of beating which passes

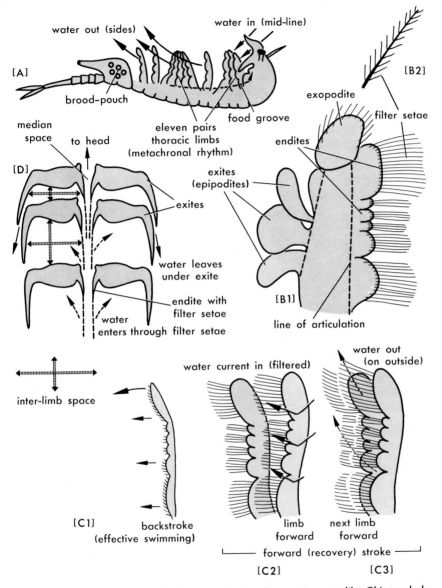

Figure 5·2. The archetypic feeding mechanism in anostracans like *Chirocephalus* and *Artemia*. **A:** General body form, swimming posture, and water currents. **B:** One of the uniform series of thoracic limbs, with endites and exites arranged to hinge backward as valve flaps. **C1, C2,** and **C3:** Propulsive and recovery strokes of thoracic limbs as viewed from the mid-line. **D:** The same recovery strokes viewed in horizontal section, showing the sequence of enlargement of the inter-limb spaces (water inflow from mid-line through endite setae) followed by compression (expelling water posteriorly under the exite flaps), which sequence results from the metachronal rhythm. For further discussion, see text.

along the series of segments. In general, water is drawn in in the mid-line and expelled posteriorly and laterally. Meanwhile a food bolus is formed in the mid-line and pushed forward. The metachronal beat alternately enlarges and reduces the boxes formed between adjacent limbs. These boxes are created during the recovery stroke because of the manner in which the various flaps (exites, exopodite, and endites) are hinged to the main part of each limb. Therefore, as is shown in Figure 5·2D, when each limb moves forward toward the head, the exites are pressed back on the limb behind by the flow of water as the animal swims along, and this prevents water from entering from the outside edge. Thus water is drawn into the temporarily expanding inter-limb space by way of the mid-line, passing from the median space through the filtering setae on the endites into the inter-limb space. However, when the next limb in series presses forward in turn, water *can* be pushed *out* under the exites and add slightly to propulsion. It should be noted that the backstroke is, of course, the effective swimming stroke with all the "flaps" nearly in the same plane as the rest of the appendage, and that this stroke does not contribute to the collection and filtration of food.

The food particles are filtered on the endite setae, and thus come to lie in the median space (*never* in the inter-limb spaces, as is suggested in some textbooks). The labrum secretes mucus which helps consolidate the food-string from the median space into a bolus between the maxillae, whence it is manipulated by chewing movements of the mandible into the mouth. It is worth stressing that in this feeding mechanism, as found in forms like *Artemia* and *Chirocephalus,* there are no specialized limbs concerned only in water propulsion or as filters. It is merely the dual action of the flaps as valves when the limb is being pushed forward in its recovery stroke that allows a series of exactly similar limbs to form a series of "temporary filtering boxes." Due to the metachronal rhythm, each inter-limb space is enlarged in turn, water is drawn in from the mid-line to a region of reduced pressure temporarily produced, and that water is filtered on the setae of the endites. While this proceeds, blood is oxygenated within these limbs and the animal is propelled forward. This mode of action in a series of like phyllopodous limbs, allowing them to subserve the functions of feeding, respiration, and locomotion, probably represents the archetypic mode of action of crustacean limbs. It is tempting to adopt the hypothesis that the limbs of trilobites were moved in a similar fashion, that is, that the rhythm of beat of their appendages was metachronal.

Figure 5·3. A: Front view of head structures in a male anostracan. The second antennae are modified for copulation. **B:** Diagram of a typical conchostracan with one valve of the carapace removed. **C:** Diagram of a typical ostracod with one valve of the carapace removed. **D:** Thoracic limb from *Argulus,* a branchiuran ectoparasite.

In some of their other characters the anostracans are somewhat more specialized. Many of them produce eggs which show a remarkable capacity to resist desiccation, and some, like those of *Artemia,* can be stored in diapause in a dry bottle on a laboratory shelf until needed. The sexes are separate and may show considerable dimorphism. The females have a brood-pouch carried on the last thoracic segment, and there are no abdominal appendages. The head of a male anostracan, shown in Figure 5·3A, demonstrates some of the more peculiar features of anostracan organization, including the stalked eyes characteristic of higher Crustacea—the naupliar eye retained in the adult suggesting

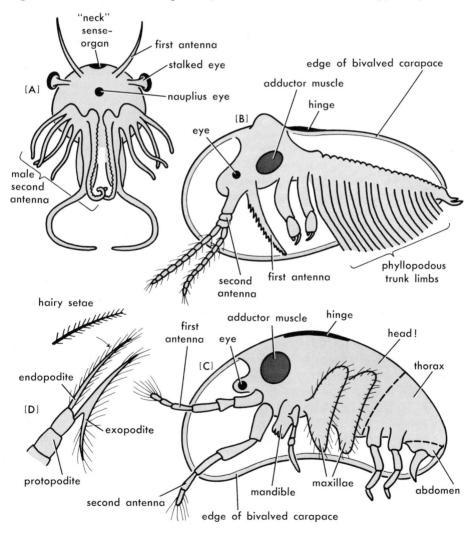

primitive organization—and the peculiar structures of the male second antennae which are used to attach it to the female as it swims in copulation.

There are only two essentially similar genera, *Triops* (known in some older books as *Apus*) and *Lepidurus,* in the order Notostraca, and they are somewhat rare, occurring irregularly in temporary ponds all over the world. The carapace forms a dorsal shield which covers about half of the body, and results in their common name of tadpole shrimps. There are numerous similar phyllopodous appendages used in swimming, feeding, and respiration. *Triops* and its allies however move along the bottoms of ponds, stirring up mud, and are apparently capable of feeding on larger particles including detritus. All the thoracic limbs have gnathobases capable of chewing, and some of this is done between the posterior limbs before the food passes forward. There is some variability in numbers of segments and limbs, which is presumably a primitive feature, but there are usually from two to five pairs of similar phyllopodous limbs on the abdominal segments and then five limbless abdominal segments to the telson. As already noted, this gives various species of *Triops* about 45 segments in all. The undifferentiated limbs are closely similar in form to those of anostracans. However, the first and second trunk appendages have modified endites of sensory importance and act as additional antennae while the eleventh in females bears an egg-cup in which the developing eggs are retained. Most populations which have been discovered consist of parthenogenetic females, males only rarely being found. In summary, notostracans can be characterized as branchiopods somewhat specialized, in limb patterns and in reproduction, for life on the bottoms of temporary ponds.

The order Conchostraca is another smallish group of rare species, with a characteristic large carapace, hinged dorsally, which can enclose the whole animal. An adductor muscle (see Figure 5·3B) is developed anteriorly and can close the two valves of the carapace. There are two series of similar phyllopodous limbs, differing in that the hind limbs do not have filtering setae but have strong spines which can be used like gnathobases to break up food particles. All species use the hind limbs of the series to chew up larger food particles, while using the front limbs for filter-feeding as in the Anostraca. An interesting reflex involves the clawed telson which, if a large particle gets jammed in the medial space, is bent forward and then rapidly straightened ventrally to kick the offending material out. In some species, the trunk limbs are no longer concerned with locomotion and the second antennae are used to produce jerky movements rather like some Cladocera. In some of their features, the conchostracans can be regarded as intermediate between the primitive branchiopod types (Anostraca and Notostraca) and the most specialized group—the highly successful order Cladocera.

More than half the eight hundred living species of the Branchiopoda are cladocerans—the so-called water-fleas. The majority are filter-feeding forms like *Daphnia,* but there are a few predaceous carnivores which catch and feed upon other small crustaceans. The group is mainly freshwater with a few marine species. A carapace covers all the trunk, but not the head, and the second antennae are the main means of locomotion producing the series of jumps which gives the group its common name. There are usually only five pairs of trunk limbs (though some forms have four pairs or six pairs), and these form one functional filter pump. The abdomen is usually reduced and bears a clawlike telson. The great majority have direct development from big eggs with the larval stages omitted. The feeding mechanism, in typical forms like *Daphnia,* is totally unlike that in anostracans and can be best understood with reference to Figure 5·4. The single nearly watertight box of variable volume forms a very fast-acting pump filter. The box is closed on the dorsal side by the ventral groove of the trunk, laterally by the folds of the carapace, has an anterior wall formed of the protopodites of legs 1 and 2, a ventral wall formed by the exopodites and bristles of legs 3 and 4, and a posterior wall formed by the fifth pair of appendages. The third, fourth, and fifth pairs of appendages are those which move, alternately enlarging and reducing the volume of the box. Their forward stroke enlarges the volume and food-bearing water is sucked in as in phase I (see Figure 5·4B1) between legs 3 and 4. As the back stroke begins (phase II in Figure 5·4B2), these close together and the water is sieved through the interlaced hairs of their bristles. The volume is further reduced and the debris squeezed on the bristles toward the food groove in phase III, and finally the fifth legs turn from the transverse to the sagittal position and water is expelled posteriorly through the median cleft. This is phase IV of the feeding process. Meanwhile, the filtered detritus pressed up into the food groove is carried forward partly by the water current and partly by mechanical sweeping from the bristles. It is kneaded to form a bolus between the mandibles and then ingested. The rapid pumping of this filtering box forms a very efficient food-gathering mechanism, so that *Daphnia* can be cultured with only bacteria as food. Although the individual limbs in cladocerans are not much modified from the branchiopod pattern, the functional pattern of their movements has been considerably modified to produce the sequence outlined above.

Some other aspects of cladoceran functional organization should be noted. The sense-organs are relatively well-developed, all forms having chemoreceptors, mechanoreceptors, and eyes. The nauplius median eye is retained in the adults of some species, but there is also a large internal eye formed of a fused pair of compound eyes. This is said to be sessile but is moved by internal musculature. It is much larger and more complex in the predaceous forms of cladocerans. For example, in *Podon*, a marine predaceous form, one-third of the bulk of the body is made up by this compound eye. The excretory organs are maxillary glands.

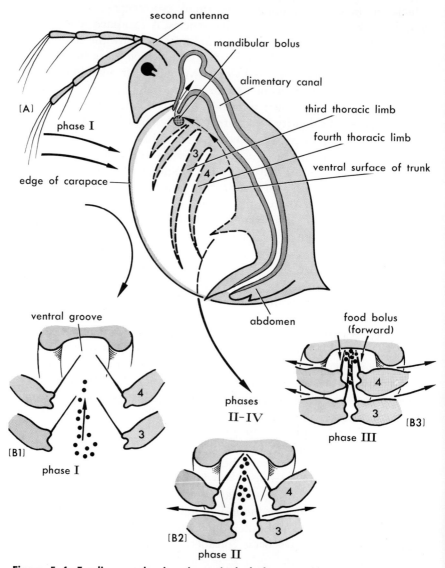

Figure 5·4. **Feeding mechanism in typical cladocerans** like *Daphnia*, where a single "box" of variable volume forms a fast-acting pump filter. **A:** General body form, swimming posture, and water currents. **B1:** Box enlarging as third and fourth thoracic limbs move ventrally (away from trunk) drawing in water and food particles in phase I. **B2** and **B3:** Box being compressed in phases II and III by third and fourth thoracic limbs in active "propulsive" stroke toward trunk, with filtered food particles being pressed into median groove. For further explanation, see text.

The heart is modified from the primitive tube of the other branchiopods, being shortened to a barrel-shape with two ostia. Since there are no blood vessels, the beating of the heart simply stirs the blood around the hemocoel. A number of freshwater cladocerans have hemoglobin as a blood pigment (or the potentiality of developing it under certain circumstances).

Sexual reproduction in cladocerans exhibits certain bizarre features. In general in freshwater populations, relatively few males are found, and parthenogenesis is common. In many cases the annual reproductive cycle appears to involve an alternation of parthenogenesis with ordinary sexual reproduction. The ecological aspect of this is that parthenogenesis can fill the pond with individuals under favorable conditions, while fertilized eggs are desiccation-resistant and can persist through winter or through drought conditions. Such resistant eggs are also of enormous importance as a factor in the passive distribution of these forms. In a few species, all fertilized eggs seem to hatch as females, and it can be shown that selective abortion during sperm maturation is responsible. The chromosome number remains constant, of course, throughout parthenogenesis, but when the unfavorable season approaches, among the parthenogenetically produced young are a few males. How this is achieved is still really unknown, but its adaptive significance is obvious. Simultaneously, eggs which need to be fertilized are produced by a reduction division. It has been suggested that substances are ingested from algal cells (which are already slowing their reproduction rate) and can induce the production of males and of meiotic eggs. As in several other groups of freshwater invertebrates, including rotifers and ectoprocts, a modification of reproductive life-cycle allows the cladocerans to exploit the peculiar conditions of the environment provided by temperate fresh waters.

Although the largest number of cladoceran genera are relatively cosmopolitan forms, similar to and feeding like *Daphnia,* there are also predaceous genera. *Polyphemus* and *Leptodora* in fresh waters, and *Podon* and *Evadne* in the sea, are specialized carnivores with huge eyes, a reduced carapace, and reduced thoracic appendages. Of course, such carnivorous forms are never present in fresh waters in such enormous numbers as the species on which they feed. Such typical micro-herbivore genera as *Daphnia* and *Bosmina* can be the principal constituents of the animal plankton of fresh waters.

A typical nauplius larva occurs in the development of some members of all groups of the branchiopods, indeed, in some species in almost all groups of crustaceans. As already noted, the nauplius hatches from the egg and then there are molt stages with the nauplius form and three pairs of appendages, and then a molt to the early metanauplius with four pairs of appendages, and only after this do subsequent molts take the development into different patterns in different groups. Later development, however, always involves the addition of other segments posteriorly (see p. 15). In several groups of crustaceans, a nauplius occurs in development but does not live a free-living planktonic existence, but is retained in a brood-pouch or even inside an egg membrane.

Of the six remaining crustacean subclasses, three groups have been enormously successful: the copepods, the barnacles, and the malacostracans, or "higher" Crustacea. The other three minor groups are treated together here merely for convenience, and not to reflect any presumed relationships.

Three Minor Groups

The ostracods form a relatively stereotyped subclass. Conservative systematics would include only one order, the Ostracoda, with a small number of genera to encompass the two hundred or so cosmopolitan species, which are not uncommon in all kinds of fresh and marine waters. The development of the carapace into a bivalve shell with an adductor muscle parallels the case of the conchostracans, but in the ostracods only the rather stenopodous head appendages are important (Figure 5·3C). Some forms retain two pairs of trunk limbs, but others have none, and locomotion and feeding are principally carried out by the head appendages. Some aspects of their physiology, particularly of circulation and respiration, and even of their comparative internal anatomy, remain obscure. Many have two pairs of excretory organs as adults, antennary and maxillary, while a few have a third set associated with the first maxillipedal segment. They occur in a wide variety of marine and freshwater habitats: some in the plankton, some burrowing in mud with spadelike antennae, some climbing weeds with prehensile antennae, some ectoparasitic on higher crustaceans with the antennae modified as suckers, and even some species limited apparently to the pools of water in certain epiphytes or to other specific plants in the tropical rainforest. Their ecology has been little studied. A number of species are luminescent, mostly belonging to three fairly common marine genera. A grotesque claim to distinction is the possession in some forms of the largest sperms in the animal kingdom. In one species of *Pontocypris,* adult males 0.3 millimeter long have been reported to have sperm 6 millimeters in length.

The minor subclass Branchiura again consists of a single order and only a few genera, of which *Argulus* is typical. The seventy-five species of so-called carp-lice are ectoparasites of both marine and freshwater fish. They exhibit a number of features clearly associated with parasitism, including: suckers on the maxillae, a piercing spine in front of a suctorial proboscis, and marked dorsoventral flattening. However, they attach temporarily only, actively swimming from host to host; they have separate and equal-sized sexes; and they lay relatively large eggs singly on stones. Their relationships with other groups of crustaceans remain obscure, but most evidence suggests that they evolved (entirely independently of other crustacean parasites) from a fairly highly evolved free-living stock. Their pair of compound eyes and maxillipeds resemble those of the higher crustaceans, while the remainder of their thoracic limbs are biramous and stenopodous with hairy spines (Figure 5·3D), somewhat intermediate in character between those of the Copepoda and the barnacles. Of course, in the ectoparasitic branchiurans these limbs are used only for locomotion and have no feeding significance.

The third minor subclass is the Mystacocarida, which was set up recently for the unique genus, *Derocheilocaris,* the first species of which was discovered and described in 1943 by R. W. Pennak and D. J. Zinn—the first specimens being found in intertidal sand at Nobska Beach on Cape Cod. Mystacocarids are minute crustaceans (about 0.4 millimeters long as adults), living in the interstitial water between sand grains on certain littorals, and have now been found in many parts of the world. Some anatomical features can be termed primitive, including the unspecialized biramous mandibles in the adult, the retention of a nauplius eye, and the retention of both antennal and maxillary glands as excretory organs. Other features—such as the long body of similar segments not grouped into tagmata—could involve the retention of ancient characteristics, or could represent specializations, along with the minute size, for life in the interstitial habitat. Most authorities would agree that the mystacocarids are in several ways intermediate between the branchiopods and the big specialized group—the class Copepoda.

Copepod Success

The subclass Copepoda is of major importance, involving more than five thousand species in the sea and in fresh waters: many of them as members of the permanent zooplankton, a few specialized for other habitats, and a large number modified as parasitic species. The free-living species—particularly those of the plankton—are relatively uniform in their functional anatomy. Ecologically, they are of supreme importance in the economics of the oceans and of larger bodies of fresh water.

Marine copepods like *Calanus* and *Temora* are probably the most abundant animals in the world. They are microherbivores (ecologically of "the second trophic level") feeding directly on the world's largest crop of primary producers—the flagellates, diatoms, and other single-celled green plants of the marine phytoplankton—a crop at least five times larger annually than that of all land vegetation, including human crops, added together (Figure 5·7). The single most abundant species —in terms both of numbers of individuals and of fraction of total animal biomass—is almost certainly *Calanus finmarchicus* or a closely similar form. Ecologically, such copepods provide the major part of the direct food supply of the world's largest animals—whalebone whales like the blue whale; some of the most abundant fishes—herring, men-

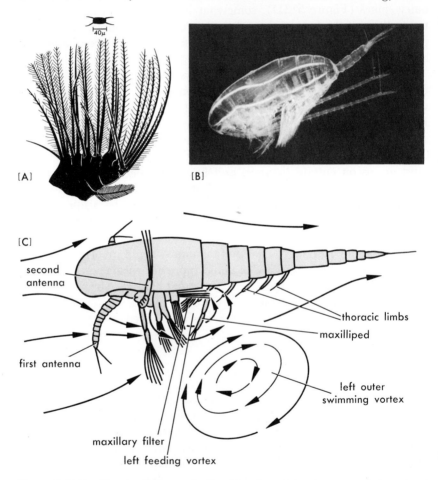

Figure 5·5. Feeding in *Calanus*. A: The filter formed by the setae of the second maxilla of *Calanus* with a food diatom, *Chaetoceros*, drawn to the same scale. **B:** Lateral view of a living *Calanus*. (See also cover photo.) **C:** Diagram of the vortices created in swimming which bring food particles into the mid-line and through the maxillary filter. [**A:** adapted from S. M. Marshall and A. P. Orr in *J. Mar. Biol. Assoc. U.K.*, 35:587–603, 1956. **B:** Photo © Douglas P. Wilson. **C:** Adapted from H. Graham Cannon in *Brit. J. Exp. Biol.*, 6:131–144, 1928.]

haden, sardines, and mackerel; and the largest fishes such as the basking sharks (*Cetorhinus*) and whale sharks. From a human standpoint,
copepods may hold one key to our future survival. As a species, our
numbers are rapidly increasing, but nearly 50 per cent of individual
humans alive at present(1968) have insufficient diets, involving short-
age of proteins. Marine copepods like *Calanus* constitute the world's
largest stock of living animal proteins. Only a tiny fraction of this is
now involved in human nutrition, and that mostly by indirect and
energetically inefficient means. (For example, calanoid copepods of
the Eastern seaboard of the United States are consumed by menhaden,
among other fish species. A multimillion dollar fishing and processing
enterprise catches menhaden and turns them into fish meal. This, in
turn, supplements the diet of swine and broiler-fowl by being incorporated in the feeding-stuffs of "intensive" agriculture. Man eventually
ingests less than 1/10,000th part of the original protein.)

The body form in the planktonic and other free-living species is relatively uniform: the elongate body without a carapace being never more
than a few millimeters in length. A nauplius eye is retained and slightly
elaborated as the single central eye of the adult, the mouth parts and
first maxilliped are used in feeding, and the other five pairs of thoracic
limbs are used for swimming. The stages in the life-cycle are always
similar, even in parasitic forms (see Figure 5·6D, E, and F). There is
a typical nauplius followed by a metanauplius with four pairs of appendages, followed by a series of copepodite stages resembling the adult
but with only three pairs of biramous thoracic limbs, and finally molting to the adult copepod. The fact that the early copepodite stages have
the posterior appendages and even the posterior segmentation underde-
veloped is of course highly significant as evidence for the pattern of
metameric morphogenesis earlier claimed as archetypic (see p. 15).
The molt to the adult usually involves the addition of two more pairs of
trunk limbs, reduction of the second antennae and great development
of the antennules, and the development of long jointed rami on the
furca of the telson. The genital openings are developed on the seventh
trunk segment behind the last pair of limbs. The exact pattern of the
adult varies in the three main orders (see Figure 5·6A, B, and C), the
distinctions depending on where the flexible joint of the trunk occurs,
and the biramous or uniramous nature of certain appendages. Separation of orders on such trivial anatomical features once again illustrates
the stereotyped morphology (among a large number of species) which
occurs when a successful pattern of animal machine is successfully
exploited.

Locomotory and feeding functions are carried out in a similar fashion in the free-living forms of all three orders. They can swim smoothly
by rhythmic beating of the trunk limbs, or, particularly in calanoids,
by jerks of their large antennules. Most of them also use a sudden

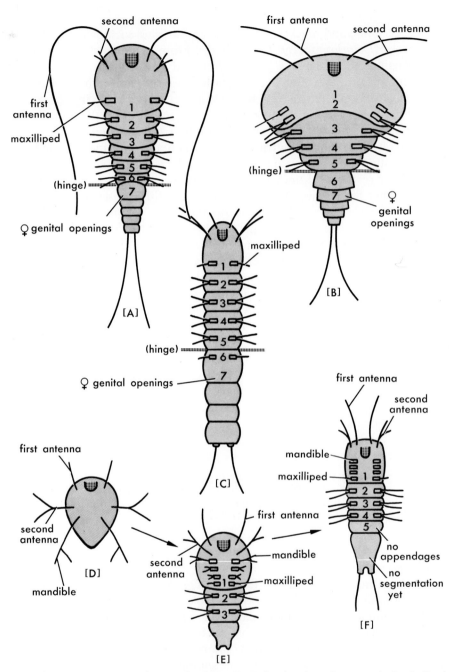

Figure 5·6. Copepod organization. A: Stylized calanoid copepod. **B:** Stylized cyclopoid copepod. **C:** Stylized harpacticoid copepod. Note the position of the flexible trunk joint in, and the minor limb differences between, the three groups. (Mouthpart appendages are omitted.) **D, E,** and **F:** Stages of larval development in copepods. (Compare with Figure 2·3.) **D:** Stylized nauplius larva (basically cyclopoid) with

flexing of the body joint as an avoiding reaction. The thoracic appendages also serve as respiratory surfaces and create their own respiratory current. In several genera like *Cyclops*, the thoracic appendages are united transversely so that the four rami of each pair beat as one. Copepods can feed by seizing individual large particles or organisms in their mouth parts, but also use a filtering mechanism. Like all other aspects of function in *Calanus*, feeding has been most thoroughly investigated by A. P. Orr and Sheina M. Marshall at Millport in Scotland. When *Calanus* is swimming slowly, a series of vortices are created in the water. These bring small particles in toward the filter mechanism, which is of the single box type (but quite unlike that in *Daphnia*). Here the box is formed of the head appendages: the mandibular palps, maxillules, maxillae, and maxillipeds all being involved. Water is drawn in by a lowered pressure created by an outward swing of the maxillipeds with their fringe of long setae. The volume of the box is then decreased and the water expelled by the maxillules being pushed forward, the expelled water passing through the sieve of the maxillary setae (see Figure 5·5A).

The particles thus filtered are removed from the "inside" of the sieve by being brushed forward by specialized long setae of the maxillipeds or by the endites of the maxillules. This brushing directs them between the mandibles and thence into the mouth. In most forms, the basketlike sieve of setae and the maxillae themselves are not moved during the filter-feeding process. However, in some species of *Acartia* the setae of the maxillae can be spread apart and then drawn back together.

As developed in calanoid copepods, the filter-feeding mechanism is superbly efficient for the retention of food organisms in the size range of from 10 to 30 microns. The same copepods also individually catch and ingest diatoms of considerably greater size. Almost all the dominant forms of the phytoplankton can be utilized as food by copepods like *Calanus*. Most investigators agree that the suspension feeding is probably a secondary development within the group, copepods in general being primarily raptorial.

As already emphasized, the calanoid copepods are dominant members of the marine zooplankton (Figure 5·7), and *Diaptomus* species often occupy a similar position in the economy of large freshwater lakes. Cyclopoid copepods are common in bodies of fresh water of all sizes and include both benthic and planktonic forms. Harpacticoid

three pairs of limbs. **E:** Stylized late metanauplius with eight pairs of appendages including two pairs of thoracic limbs (basically calanoid, workers on zooplankton would designate this stage "nauplius VI"). **F:** Stylized copepodite stage with three pairs of thoracic appendages and the fifth thoracic segment already differentiated.

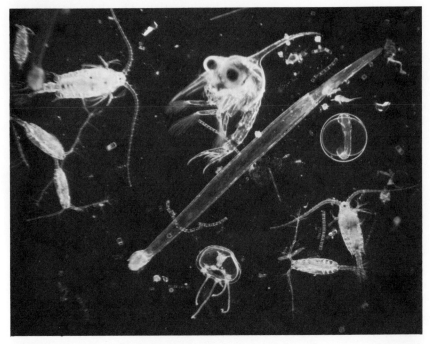

Figure 5·7. Living zooplankton. Various calanoid copepods are seen along with an arrow-worm, *Sagitta*, diagonally across; a zoea larva of a crab above the arrow-worm, near center; a developing fish egg, below and to the right; a medusa, near the bottom, the fine chains of diatoms which serve as the basic food supply for all these forms of the animal plankton. [Photo © Douglas P. Wilson.]

copepods are mostly benthonic, living in and over bottom-deposits in both the sea and fresh waters. A few relatively unspecialized ectoparasitic forms occur in the latter two orders, and it may be from forms like these that the more specialized parasitic orders have evolved.

Copepod Parasites

There are numerous parasitic genera in the subclass Copepoda, and they can be ranked as a series showing progressively greater degrees of modification for the parasitic habit. They range from commensal scavengers which can be occasional parasites through ecto- and endoparasites to extremely modified tissue parasites.

The simplest anatomical modifications may be those of the mouth: the labrum and labium forming a sucking tube around the mandibles which become elongate piercing stylets. In more specialized forms there is a reduction of appendages and attachment to one host is permanent. In some, all parts of the attached adult become reduced except the gonads, and in many of these the males are shrunken, tiny forms permanently attached to the females. In the most modified forms, only the fertilized female survives as a bag of tissue—which is partly host tissue produced by reaction—enclosing the mature female gonads which enter a continuous and long-continued egg production. It is noteworthy that almost all species, even the most modified, show the entire series of larval stages: from nauplius to metanauplius to copepodite to adult copepod (see Figure 5·6D, E, and F).

The least modified copepod parasites fall into two groups. First, those where only the mouth is slightly modified and the free-living organization of body and appendages is retained: such forms live in the lumen of the intestine of fishes, marine mammals, and large decapod crustaceans. Most of the species in such genera as *Botachus* and *Enterocola* are host-specific, although they usually only feed on faeces and rarely graze on the intestinal wall. Secondly, there is a group of ectoparasites of fishes, like *Ergasilus* (see Figure 5·8A) and allied genera, which feed mainly on mucous secretions and have the antennae modified as clutching hooks. Related forms live on the gills and in the mouth cavity of marine and freshwater fishes and suck blood from gill lamellae and other exposed tissues.

Caligus (see Figure 5·8B) and its allies show the further modification of the maxillae developed as suckers, grasping mandibles, and a generally clumsy build with an enlarged genital segment. These forms can still swim from host to host and are specific ectoparasites on the gills of various fishes and of *Nautilus*. The adult females of *Chondrocanthus* (see Figure 5·8C) cannot swim and are permanently attached to the host after settling there. The appendages are reduced to blunt lobes, and the minute male is permanently attached to the genital segment of the female. The egg-sacs are in the form of a pair of long continuously produced filaments. Species of *Chondrocanthus* are permanent parasites on the gills of various fishes.

Both life-cycle and anatomical specializations are a little more complex in species of the genus *Lernaea*. In the copepodite stage, they are temporary parasites on fish gills, leaving the first host, which is usually a flatfish, after molting to adult copepods. They then copulate and males die, but the females attach to gadid fish, usually in the gill tissue. The body becomes vermiform with an enormously enlarged genital segment, and the anterior part of the head modified into a series of branching roots which grow deep into the musculature of the fish (Figure 5·8D). Once again, there is continuous egg production for a long period of time. *Xenocoeloma* is a genus of parasites of polychaete worms, the adults forming mere sacs, feeding by roots completely within the host. There is an extraordinary degree of host-collaboration, the sac wall being largely of host tissue, with even an invagination of

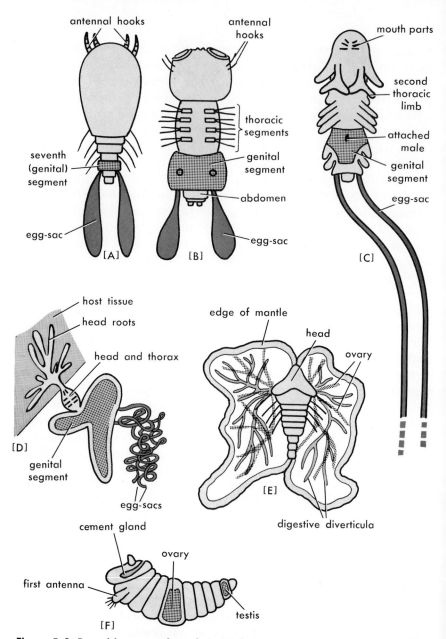

Figure 5·8. Parasitic copepods and cirripedes. A: *Ergasilus,* a copepod ectoparasitic on fishes. **B:** *Caligus,* another fish ectoparasite, with maxillary suckers and an enlarged genital segment, but retaining a basically copepod body form. **C:** *Chondrocanthus,* a permanently attached copepod parasite of fishes, with a minute male and continuous egg production. **D:** *Lernaea,* adult female copepod parasite, with branching roots embedded in the host fish's musculature. **E:** *Laura,* an adult parasitic cirripede whose host is an antipatharian coelenterate. **F:** *Proteolepas,* an unusual maggotlike cirripede, discovered by Charles Darwin as a parasite of stalked barnacles.

the host coelom to provide a sort of false alimentary canal for the mass of parasite gonads. A normal nauplius is known, but some details of the life-cycle are controversial. Apart from the two long external egg-sacs, the adult form of *Xenocoeloma* is completely unlike a copepod.

In one group—*Monstrilla* and related forms—the late larvae are parasitic and the adults free-living. Nauplius and metanauplius are free-living, but after this the host worm or mollusc is entered. The copepodite stages are passed as a long wormy parasite which builds up food stores living in the blood vessels of the host. It then metamorphoses and breaks out as a relatively unmodified adult copepod lacking, however, any functional gut. The adult lives on its stored food and produces eggs, and thus acts as the dispersal stage in these parasites.

Barnacles

The major subclass Cirripedia is a widespread and successful marine group of nearly one thousand species. The majority are typical barnacles, specialized as adults for an entirely sessile mode of life, but there are also a series of parasitic forms showing progressively more extensive modifications. All have similar life histories. In characteristic barnacles (see Figure 5·9)—all placed in the order Thoracica—the head becomes fixed down to the substratum by the antennules at metamorphosis, and the adult lives permanently attached using six pairs of biramous, thoracic limbs as a food-catching mechanism, which kick food into the mouth. The abdomen is reduced to a vestige, and most forms have become hermaphrodite.

Cirripede nauplii have characteristic frontal horns and a dorsal spine (see Figure 5·10A), which are retained through a series of molts to the last metanauplius. Then the mouth parts are reduced and the only head appendages are the antennules with their terminal suckers. A carapace forms a bivalve shell around this cypris-stage larvae which has six pairs of similar trunk limbs and a reduced abdomen. The cypris settles and then metamorphoses into the adult barnacle. The first attachment of settlement is by the antennules and then cement glands secrete while metamorphosis takes place by a twisting of the trunk (see Figure 5·9B), the mouth with eye and adductor muscle moving away from the antennule. A fleshy mantle then replaces the cyprid carapace, and begins to secrete calcareous plates. There are two major patterns of further development in the Thoracica: either toward a stalked barnacle like *Lepas* where the fleshy stalk supports a head with calcified plates, or toward an acorn barnacle like *Balanus,* where the conical limpetlike body is totally enclosed in calcareous plates secreted by the mantle. After the shell plates of the adult have been formed, they grow more or less continuously and totally independently of the molt cycle. In

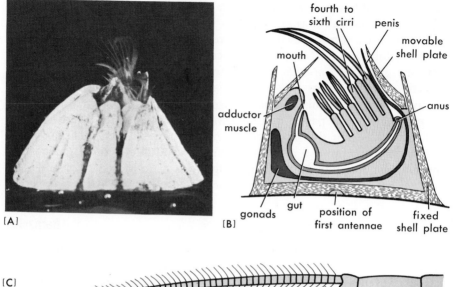

[A]

[B]

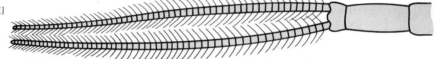

[C]

Figure 5·9. Functional morphology of acorn barnacles. A: *Balanus crenatus,* a typical acorn barnacle photographed alive while "fishing." **B:** The general anatomy of an adult acorn barnacle, showing the fourth, fifth, and sixth thoracic limbs as the elongate cirri used in food collection. **C:** A single cirrus, biramous, many-segmented, and fringed with stiff setae. For discussion of feeding methods, see text. [A: Photo © Douglas P. Wilson.]

barnacles, molts really only affect the exoskeleton of the cirri, as the modified trunk limbs are called, and the surfaces associated with them. There is a very constant pattern of shell plates throughout the Thoracica, and we can trace evolution involving reduction of the number of plates in several lines.

Barnacles are a very successful group of sessile animals, dominating particular zones of the rocky littoral in most parts of the world and incorporating vast numbers of individuals in some species. They have the distinction that Charles Darwin worked for a number of years on their systematics and distribution. Over the last twenty years, there has been a great revival of interest in barnacles—particularly in aspects of their physiology and population ecology. This recent interest is partly "spin-off" from intensive applied research carried out during the Second World War. Research teams in both the United States and Britain were involved in preventative work on ship fouling. Barnacles are among the most important of marine fouling organisms, and, if not prevented, can add more than 15 per cent to the fuel consumption of an average-sized merchant vessel, or decrease by several knots the maximum speed of a warship.

Most species of both stalked and acorn barnacles are hermaphrodite. In relation to the sessile habit, many have long penes and cross-fertilization is achieved by their extension through the colony. A few forms liberate sperms which are taken in by the feeding mechanism to cross-fertilize. Some species of the stalked barnacle *Scalpellum* have dwarf, semiparasitic males attached to the mantle of the large females.

All barnacles catch their food particles with their cirri, three or six pairs being involved, but there are a number of different patterns of feeding activity. In all forms, each cirrus has two long, many-segmented rami, bearing stiff bristlelike setae. The only other appendages in the adult are the relatively tiny mouth parts. The methods of feeding which have been described can be classified into five main patterns, and a few species of barnacles can employ three of them. The first method utilizes the cirri as a passive net, deliberately turned to be held across the direction of any water current. The more posterior and longer cirri (three pairs in *Balanus* and its allies) are those used to form the net and they may remain extended for periods of minutes before being withdrawn. Extension of the cirri is always relatively slow since it is due to a hydraulic mechanism, blood being pumped from the rest of the body into the lumina of the cirri by muscular action. Cirral withdrawal is rapid since it is performed by the flexor muscles which both roll up the cirri in an oral direction and pull them into the "mantle-cavity." The second pattern of feeding activity uses rhythmic movements of the cirri as a sweep net. The long cirri are extended, swept forward and downward in a scooping action, and withdrawn, each cycle taking about half a second. A third type of feeding behavior involves faster repetitions of the first part of the cycle, the cirri being used again and again in sweeping movements, only being fully withdrawn at long intervals for the ingestion of the trapped food particles. A fourth feeding mechanism involves the short cirri which are not normally protruded from the shell. Water currents pass through these each time the barnacle opens and filtration of fine particles can occur even in the absence of activity of the long cirri. These four filtering mechanisms can apparently collect suitable food organisms ranging in size from 2 microns to over 800 microns in length. A fifth feeding mechanism—exhibited principally by *Lepas* and its allies—is not really filter-feeding. In these forms an individual cirrus may catch a small animal, for example a copepod, and it then rolls down, independently of the other cirri, and passes the captured organism to the mouth parts. Local chemoreception is involved in such capture.

The functional systems of barnacles are relatively stereotyped. There is a U-shaped gut (see Figure 5·9B) with an expansion forming a

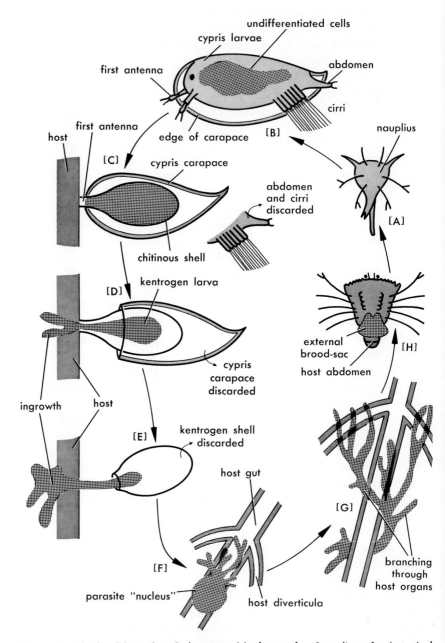

Figure 5·10. The life-cycle of the parasitic barnacle, *Sacculina.* **A:** A typical barnacle nauplius is released from the brood-sac. **B:** A typical barnacle cypris larva, though lacking a functional gut, is the stage which follows the metanauplius. **C:** The cypris attaches to a host crab by the first antennae, and discards the cyprid appendages. **D:** A further molt of the cypris carapace gives rise to the kentrogen larva, as the undifferentiated parasite tissues begin to invade the host. **E:** The chitinous shell

simple stomach. There are no blood vessels or hearts, the body fluids being stirred around by the movements of the cirri and mantle, which organs also provide the respiratory surfaces. The nervous system shows some concentration in the trunk region but is relatively simple in organization. The adult excretory organs are maxillary glands.

Cirripede Parasites

As with the copepods, parasitic barnacles show a series of degrees of specialization. In this subclass, the parasitic habit probably first arose from the fact that many regular barnacles in the order Thoracica attach themselves as fouling organisms to other animals, for example to big molluscs and decapod crustaceans, even to whales and turtles, and are thus transported about. Supplementing their nutrition by sending some roots into the host tissue was probably the first stage of ectoparasitism in such barnacles. Almost all parasitic cirripedes retain a typical barnacle life history from horned nauplius through cypris stages, and it is usually the cypris larva which first attaches to the host organism.

In the order Acrothoracica, the least modified species live as adults as ectoparasites on whales, turtles, and sharks, retaining a rough acorn-barnacle shape, but deriving food through a root system in the host tissue. Others live as ectoparasites in the gill cavities of lobsters and crabs, while others bore into the shells of molluscs. A few allied species bore, by an unknown method, into coral or limestone rock and continue to use their cirri in feeding. The order Ascothoracica consists of species parasitic on coelenterates and echinoderms, which, though mostly embedded in the host tissues, often retain fairly obvious cirripede organization. Some show the head structures, six pairs of thoracic appendages, and a segmented abdomen, but in all the mantle is enormously developed as an absorptive sac containing diverticula of the parasite's gut and branches of ovarian tissue. The males are again minute. The genera *Synagoga* and *Laura* (see Figure 5·8E) are parasites on antipatharian coelenterates. The order Apoda consists of a single genus, the maggotlike *Proteolepas* (Figure 5·8F), parasitizing stalked barnacles, which was discovered and described by Charles Darwin.

Finally, the order Rhizocephala, the species of which are almost all parasitic on crabs and lobsters, have claims to being the most highly modified parasites in the animal kingdom. The adult is merely a branch-

---

of the kentrogen larva is discarded in turn, as the remainder of the parasite moves internally. **F, G:** The ingrowth of parasite tissue forms a "nucleus" on the host crab's gut and sends branches through all the visceral organs of the host. **H:** a brood-sac, containing developing eggs of the parasite, is formed externally under the host's reflexed abdomen. From this are released the nauplii of **A.**

ing structure—rather like a fungus—growing through the host tissues, and at no stage in the life history is there a functional alimentary canal. Externally, the nauplii and cyprid larvae are like those of typical cirripedes, but internally they have undifferentiated cells in place of a gut.

In *Sacculina,* the typical cirripede nauplius is retained in a brood-sac and liberated as a cyprid without a gut. This attaches to a bristle on the limb of the crab by its antennules, then moves to the thinner cuticle at an articular membrane. Its tissues bore in, and there is a loss of the cyprid exoskeleton (see Figure 5·10), then the loss of a further larval skin. The motile tissue mass is carried in the bloodstream to the midgut of the crab, where it forms a nucleus from which parasitic tissues branch out through the animal in all directions. At a subsequent molt of the host, a brood-sac is formed externally in the reflexed angle of the crab's abdomen. It has long been known that all crabs parasitized by *Sacculina* look like immature females, and this is often referred to as resulting from the destruction of the gonads by the parasite. Thus it is quoted in many textbooks as an example of parasitic castration. If a male crab is infected with *Sacculina,* then, with successive molts, it develops a female appearance: the copulatory organs and other dimorphic limbs change as does the shape of the abdomen. An infected female crab begins to look progressively less mature. The older texts suggest that these secondary sexual characters are altered when the animals are castrated by the parasites, but there has always been the difficulty that, in normal conditions, some of these secondary characters appear long before the gonads. Further, in parasitized specimens, the sex character change appears before the gonads are completely destroyed. It is now known that the androgenic gland (see p. 51) controls the development of such characters in male crabs, and it appears that that endocrine organ is affected early in parasitization.

Species of *Peltogaster* are similar tumorlike parasites living in hermit crabs. The genus *Thompsonia* on similar host species forms funguslike ramifications capable of asexual reproduction. They also form numerous egg-sac bags at any available joints in the host exoskeleton, which appear regularly at every molt. The tissue of the egg-sac wall is apparently that of the host.

Crustacean Diversity II

THE NUMEROUS FORMS making up the remaining, and by far the largest, subclass of the Crustacea, the Malacostraca, are usually called the higher Crustacea. Once again, however, the most primitive members of the group are filter-feeders with relatively undifferentiated limbs, relatively phyllopodous, and arranged in a uniform series with little secondary tagmatization.

The accepted phylogeny of the group is based on unusually extensive and detailed studies of comparative morphology, on a real understanding of archetypic function in a few critical types, and on some remarkably pertinent fossil forms. The scheme of interrelationships which has been deduced is therefore an atypically sound piece of exegesis but, unfortunately, a very complex one. Malacostracan evolution has *not* involved a single line of increasing specialization. There are at least four minor lines of "experimental" patterns, and two enormously successful lines of evolutionary specialization, parallel in some features but totally independent, both originating in rather undifferentiated filter-feeding types. Systematically, these are outlined in Table 5·1, in the two series, six divisions, and many orders and suborders of the subclass Malacostraca. As all "natural" classifications are intended to do, this attempts to portray the hierarchy of interrelationships which have been hypothesized. It is significant that, in this one, between class and order, three intermediate levels have to be invoked: subclass, series, and division (or superorder). Further, the accepted classification of certain genera of crabs involves seven more intermediate levels between order and genus: suborder, section, subsection, superfamily, family, subfamily, and tribe. The student should realize that this elaborated hierarchy does not represent mere gamesmanship of museum zoologists, but an honest attempt to portray a complex series of interrelationships involving a highly successful group (of over eighteen thousand species), which has utilized a remarkably stereotyped basic pattern of anatomy.

The segments and tagmata have a remarkably constant arrangement throughout the Malacostraca. The head has 5 overt segments (almost certainly 6, embryonically), the thorax has 8 (up to 3 of which may become associated with the head and carry maxillipeds), and the abdomen has 6 segments. The only exceptions are the species of *Nebalia* and two allied genera where there are 7 abdominal segments, and these few species form the minor series Leptostraca. All others show the constant pattern, 5 plus 8 plus 6 (see Figures 6·1 and 6·2B); throughout the Eumalacostraca there is no variation in segment numbers. Within the Eumalacostraca, the Syncarida (with thirty-two species), Thermosbaenida (with four species), and Hoplocarida (with one hundred

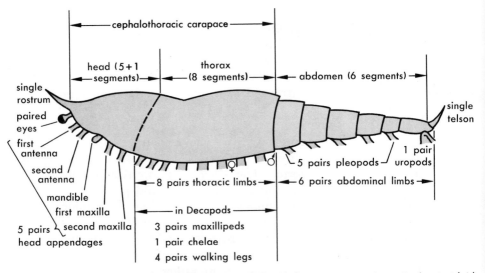
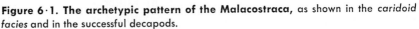

Figure 6·1. The archetypic pattern of the Malacostraca, as shown in the *caridoid facies* and in the successful decapods.

and eighty species) are small specialized groups off the major evolutionary lines. The two main stocks are the Peracarida and Eucarida, each of which numbers about nine thousand species. Within each of these divisions is a series from filter-feeding forms—the mysids in the Peracarida and the euphausiids in the Eucarida—to highly modified carnivores and scavengers. These range from forms with a uniform series of phyllopodous limbs to those like crayfish, lobsters, and crabs, with chelae and other specialized limbs. A moderately good fossil record and the existence of undifferentiated living forms like euphausiids, in which limb function can be studied, makes the construction of a convincing archetype somewhat easier than is usual. In fact, the archetypic plan of organization in the Malacostraca has long been defined, and is termed the *caridoid facies*. In this pattern, the diagnosis of the three tagmata runs: head, with five pairs of appendages (two pairs antennae, mandibles, and two pairs maxillae); thorax of 8 segments with uniform limbs, each with a walking endopodite, with an inwardly directed fringe of setae for filter-feeding, a swimming-paddle exopodite, and a respiratory epipodite; and an abdomen of 6 segments, 5 with swimming pleopods, and the sixth bearing uropods which, with the telson, form the tail fan.

Independent evolution of more specialized forms in the two groups—Peracarida and Eucarida—has resulted in parallel loss of the undifferentiated characters of the *caridoid facies*. The diagnostic features of the peracarid line include direct development, the thoracic segments not fused together and at least 4 being free from the carapace if present, and possession of a brood-pouch formed of the oöstegites on the thoracic limbs. The features of the Eucarida include a complicated larval development in most forms, a cephalothoracic carapace fused with the dorsal parts of all the thoracic segments, and no brood-pouch formed of oöstegites. Before considering these two major stocks, the primitive relicts and other aberrant forms making up the four minor groups of Malacostraca will be briefly discussed.

Experimental Patterns

In the series Leptostraca, there is a single order Nebaliacea consisting of three living genera. *Nebalia* and its allies have strong claims to being archetypic for the Malacostracan lines. Closely similar fossil genera are found more or less continuously back to the Cambrian. The eight thoracic segments bear uniform phyllopodous appendages, closely similar to those of *Hutchinsoniella* and to the limbs of trilobites. Other features are peculiar to the group: a two-valved carapace with an adductor muscle, a jointed rostrum, stalked eyes, a peculiar first antenna which may be primitive, and 7 abdominal segments ending in a telson with rami. The last features are totally unlike the other malacostracans, as is the reported occurrence in some species of a series of eight pairs of segmental organs. The young show direct development.

Leptostracans are all detritus-feeders, using the thoracic appendages in a mechanism with resemblances to—but probably not directly homologous with—that of the more primitive branchiopods. As *Nebalia* moves through the fluid mud in which it lives, a food stream is drawn in anteriorly and the water passes out at the posterior of the carapace. The setae on the podomeres of the endopodites of each limb form the filtering mechanism (see Figure 6·3A). The functional morphology in these forms is largely known as a result of the studies of H. Graham Cannon. Typical species of the group such as *Nebalia bipes,* live characteristically in soft mud under stones in the littoral in many parts of the world. The genus *Nebaliopsis* lives in the oceanic plankton. It is almost certain that the typical mud-dwelling forms were evolved with some specialization of carapace and feeding mechanism from the free-swimming animals of the *caridoid facies,* and that *Nebaliopsis* represents a subsequent return to a planktonic habit from such detritus-feeding forms. In contrast to the eighteen thousand known species of the series Eumalacostraca only seven living species of leptostracans have been described.

Undeniably, the most primitive forms included *within* the higher crustaceans or Eumalacostraca are the thirty-two species of the division Syncarida. They all live in fresh waters in rather relict habitats and show most of the diagnostic features of the *caridoid facies,* although they do not have a carapace. There is little difference between the thoracic and abdominal segments; in this, they are more primitive than all the other malacostracans, and they have no specialized grasping limbs, either chelate or subchelate. Numerous fossil forms appear to be closely related to the extant syncarids, particularly from the Carboniferous Period, and the presently living species are limited to geographically and ecologically peculiar habitats. This has probably an evolutionary significance similar to that of the distribution of present-day lungfishes. *Anaspides* is a medium-sized (reaching 5 centimeters long) shrimplike form found in mountain springs around 4,000 feet in Tasmania; *Paranaspides* and *Koonunga* are modified forms living in larger lakes in Tasmania. They have direct development from large, single eggs. *Bathynella* is minute (about 2 millimeters long) and up to the late 1950's was known only from three cave localities: one in Wales, one near Prague in Czechoslovakia, and one near Kuala Lumpur in Malaya. It has recently been rediscovered as a member of the interstitial fauna of river sands, in twelve more localities all over the world. The genus may be much more widely distributed than has been suspected, but still shows a relict distribution; restricted to underground waters and to the peculiar habitat provided by the interstitial water in certain types of sand deposits in the beds of rivers having certain physical characteristics.

The second division of the series, the Thermosbaenidae, was recently set up for two genera, *Monodella,* including two Italian cavernicolous species, and *Thermosbaena,* from hot springs in Tunisia. *Thermosbaena mirabilis* cannot survive below 35°C, and tolerates temperatures of 47°C (that is, close to the absolute upper limit for metabolism in metazoan tissues). They are clearly eumalacostracan animals, but cannot be fitted into any of the other groups.

Figure 6·2. Three types of malacostracan crustaceans. A: *Mesodopsis slabberi,*
a mysid. (Compare with Figure 6·3D.) **B:** *Palaemon serratus,* the "common prawn"
of European waters. (Compare with Figure 6·1.) **C:** *Squilla desmaresti,* a hoplocarid
or "mantis-shrimp." [Photos of living animals © Douglas P. Wilson.]

The division placed fifth and last in Table 5·1, the Hoplocarida,
consists of about one hundred and eighty species placed in *Squilla*
and related genera. These are not primitive forms, but have several
features which separate them from both major stocks of malacostracans (Figure 6·2C). They are all marine, living in burrows, with
a short carapace on 3 thoracic segments. The abdominal segments are
unusually large and well-developed, with appendages for swimming
and respiration. The head shows an anomalous division into segments,
and the first five thoracic appendages are subchelate. There is a planktonic larva of a peculiar type, the erichthus larva. In summary,
the hoplocarids, or Stomatopoda, represent a specialized rather than a
primitive group, but one which is well off the main lines of evolution
in the higher Crustacea.

Peracarid Stocks

The division Peracarida encompasses a series of six orders, ranging from the mysids which retain many features of the *caridoid facies* through some intermediate groups to the Isopoda and Amphipoda. These last two orders are highly modified, bottom-dwelling scavengers and carnivorous macrophages. The more important characteristics of the division have been noted earlier, including the direct development and the possession of an oöstegite brood-pouch, but a few others are worth detailing. Only in Mysidacaea is there a carapace attached to the first 4 thoracic segments. Three segments are involved in the Cumacea, and 2 in the Tanaidacea, while the carapace is absent in the highly successful isopods and amphipods. Throughout the division, the first thoracic segment—and it only—bears maxillipeds, while the mandibles are often asymmetric and all have an articulated movable tooth called the *lacinia mobilis*. In the limbs, if there is any tendency for reduction of the number of podomere units to occur, it is the carpopodite and protopodite which are involved. (In the eucarid line, the ischiopodite and preischiopodite fuse if any reduction occurs.) The fact that these relatively trivial anatomical characters are entirely diagnostic of the more than nine thousand species in the division is highly significant. Once again, evolutionary success involves a stereotyped pattern.

The order Mysidacea consists of a very closely knit group of about four hundred and fifty species, retaining many features of the *caridoid facies*. They occur in the marine plankton and over littoral sand-flats, and there are a few forms in fresh waters which physiological investigations have shown to be relatively recent marine relict species. Some of the species in such genera as *Mysis* and *Praunus* are often very abundant as individuals, forming extensive and dense flocks in shallow inshore waters over sand, and forming a major part of the food supply for such fish as flounders (see Figure 6·2A).

The order shows the group characters already mentioned: the single maxilliped being followed by seven pairs of undifferentiated stenopodous, but bristly, biramous appendages. The pleopods are small, except for the last pair, which form the uropods of the tail fan (with the telson). Statocysts in these uropods betray their use as both rudders and elevators in steering the swiftly darting mysids. Mysids are excellent swimmers with two entirely different feeding mechanisms. First, largish particles of food can be seized by the endopodites of the thoracic limbs and placed between the mandibles for ingestion. Secondly, the second maxillae (Figure 6·3B) form a flat floor below the mouth and, along with the maxillule and maxilliped, act as a filter press—mechanically similar to that of *Daphnia,* though involving different appendages. In *Hemimysis* and related forms, each thoracic limb rotates, causing a vortex in the water which draws in particles to the mid-line (Figure 6·3D), whence they are transferred to the maxillary filter press (Figure 6·3C). In other forms the filter apparently receives only particles stirred up from the bottom by movements of the animal. From an evolutionary viewpoint, the first feeding method is that of the higher malacostracans, while the filter-feeding resembles that of some branchiopods.

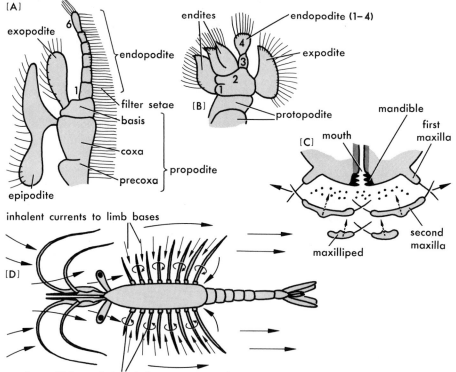

Figure 6·3. A: Thoracic limb of the leptostracan *Nebalia*. **B: Second maxilla** of a typical mysid. **C: The maxillary filter press** used in the filter-feeding of mysids in vertical cross section. **D: The feeding and locomotory currents of a mysid,** viewed from the dorsal side. (Compare with Figure 6·2A.) For further explanation, see text.

Much of the biology of mysids requires further investigation. They have excellent sense-organs like predators, are good swimmers, and a number of planktonic forms show a diurnal, vertical migration: living on the bottom by day and swimming in the surface waters by night. In spite of the many primitive features in their functional morphology, it seems possible that many mysids could live as predaceous carnivores at night.

A similar body form, but modified for life burrowing in sand, is shown by the more than five hundred species of the order Cumacea. The carapace is expanded laterally as a gill-chamber, the eyes are sessile, and the uropods filamentous. Such forms as *Diastylis* are still filter-feeders, using the setae of the maxilla. The pattern of the water current for feeding and respiration is somewhat peculiar: both inhalant and exhalant openings to the carapace-chamber being at the anterior end. Other forms including *Iphinoë* and *Cumopsis* have an unusual feeding mechanism. Essentially it involves scrubbing the organic matter off individual sand grains, using the first and second pairs of maxillipeds to tumble the grain and scrape the food material from its surface. Cumaceans show a cosmopolitan distribution, are all burrowing animals, and are found most typically in the lower littoral and sublittoral zones.

The order Tanaidacea is another smallish group (about two hundred and fifty species) of burrow-dwelling or tube-building forms, which are in some ways intermediate between cumaceans and isopods in the degree of loss of the *caridoid facies*. The carapace is very small, the eyes usually sessile, and the uropods filamentous—all these features connected with burrowing. The thoracic exopodites are absent or vestigial, while some species have chelae on the anterior thoracic limbs. Both these last intermediate orders of the peracarid stock have shown increasing development of the endopodite and reduction of the exopodite in their thoracic appendages. Thus, they are better food catchers and graspers, and increasingly poorer swimmers.

The recently described order Spelaeogriphacea, involving a single genus discovered in 1957 in caves in South Africa, may be related to the last two orders, but is more likely to prove an ancient (but physiologically specialized) offshoot of a mysidlike stock.

Peracarid Success

The two remaining advanced orders of peracarids—Isopoda and Amphipoda—are highly successful and numerous (four thousand species and three thousand six hundred species respectively). The animals in both groups are poor swimmers, and highly modified as scavenging macrophages and bottom crawlers. They have in common such features as the lack of a carapace, sessile eyes, the first pair of thoracic limbs as maxillipeds, the remaining thoracic limbs stenopodous in form and without exopodites, the pleopods respiratory in function, and uropods not forming a tail fan. Diagnostic differences between the groups are concise and consistent. The isopods are dorsoventrally flattened, with all thoracic limbs similar and all abdominal limbs similar. The amphipods are laterally compressed, with thoracic and abdominal appendages, each arranged in at least two groups, differing in form and function, so that there are from four to six distinct series of cormopodites (or trunk limbs).

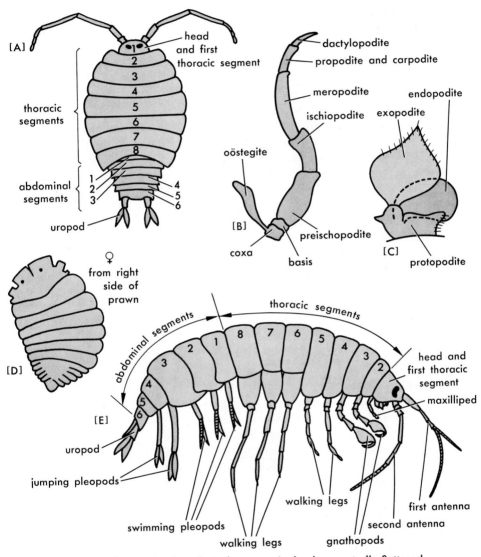

Figure 6·4. Peracarid organization. A: Body pattern in the dorsoventrally flattened *Ligia*, a typical isopod. **B:** Thoracic limb from *Ligia*, typical of isopods. **C:** Abdominal limb of same. **D:** A parasitic isopod, *Bopyrus*, from the gill-chamber of a prawn. For further discussion, see text. **E:** Lateral view of an archetypic amphipod (actually basically gammarid), with characteristic lateral flattening of the body which bears seven types of body limbs in series.

Perhaps the largest number of species in the order Isopoda fall into genera like *Ligia,* which live on the seashore as intertidal scavengers. These all show the group characteristic of having "all limbs similar": thoracic limbs with a single ramus of the endopodite for walking and foliaceous abdominal limbs for respiration. There are only one pair of maxillipeds at one end and one pair of uropods at the other, which are different from the two patterns. They have oöstegites on the bases of the thoracic limbs forming a brood-pouch, and they are usually dorso-ventrally flattened (see Figures 6·4A, B and C). The numerous species of *Ligia,* and allied genera such as *Idotea, Oniscus, Porcellio, Gaera,* and *Sphaeroma,* vary a little in shape and a great deal in size (adults range from microscopic species to forms about a foot in length), but are almost uniform in their structural organization. The only anatomical modification is a tendency for fusion of the abdominal segments.

There is a predisposition toward the terrestrial habit in several littoral forms. In many parts of the world the vertical zones of the seashore have specific isopods identified with them, the higher ones being progressively less aquatic in all aspects of their physiology, including respiration and excretion. Several genera, but notably *Porcellio* and *Armadillidium,* are completely terrestrial and are known as wood-lice or sow-bugs or slaters. The first pleopod is enlarged and extends posteriorly, covering the others which can thus be kept damp as respiratory surfaces in the air. In forms like *Armadillidium,* the cuticle on these limbs is invaginated into a ramifying system of tracheae. These obviously increase the efficiency of aerial respiration, even though the tracheae do not penetrate the body of the animal.

The development of tracheae in isopods totally unrelated to the other terrestrial arthropods is a classical example of independent evolution. Behavior in terrestrial isopods has been much studied, and shows all the taxes and kineses in response to such variables as light and temperature, as are appropriate to keep these imperfectly terrestrial animals from the dangers of water loss. Crudely, their behavior is such as to make them cryptic or nocturnal or both. A few genera of aquatic isopods have become euryhaline and invaded estuaries, and there are some completely freshwater genera including *Asellus.*

There is also a series of parasitic isopod species. The fish-louse, *Aega,* is a normal isopod in body form but with piercing mouth parts and hooks on the thoracic limbs. Adult specimens of *Bopyrus* are permanently fixed parasites living under the carapace of decapod crustaceans attached to the gills. The sex of the adult parasite is apparently not determined until after its arrival in the gill-chamber of the host. The first specimen of *Bopyrus* to invade a decapod inevitably becomes a female, and any later arrivals become males. If a young female is removed from a newly infected host and put into the gill-chamber where there is a fully adult female, then the young one's sexual development is retarded and reversed, and it will eventually become a male. Similarly, young males will develop into females if removed from the presence of females and put into the gill-chambers of uninfected shrimps or prawns. The fully developed males are smaller and rather less modified than the females. A further peculiar feature is that the mature females show an asymmetry corresponding to whichever side of their host they live on (see Figure 6·4E).

Other species of bopyrid genera are more modified tissue parasites, forming galls on the sides of various shrimps and crabs and often causing parasitic castration. An extreme case of specialization is shown by some parasitic isopods whose hosts are cirripedes of the order Rhizocephala. The isopod genus *Danalia* is hyperparasitic on *Sacculina*, while *Liriopsis* invades *Peltogaster* (parasite on hermit crabs), and these hyperparasites each castrate the parasites in turn. In the latter case, the hermit crab nourishes the funguslike growth of the cirripede parasite, which in turn nourishes the reproductive cells of the isopod hyperparasite, and, of course, the eggs which are produced in the little external egg-sacs are those of the isopod.

Figure 6·5. An amphipod "skeleton shrimp," *Caprella aequilibra,* climbing on the branches of a red seaweed, *Ceramium.* [Photo © Douglas P. Wilson.]

The enormous and successful order Amphipoda is usually characterized by a lateral flattening of the body, and almost always by division
of functions among the varied limbs. There are about three thousand
six hundred species, some of which are among the most abundant animals of the marine littoral. The gammarids—which include the abundant sand-hoppers or beach-fleas—have seven types of body limbs in
series. The first of the thoracic appendages are maxillipeds, the second
and third subchelate gnathopods, the fourth and fifth forwardly directed walking legs, the sixth, seventh, and eighth backwardly directed
walking legs. Then there follows on the abdomen: three pairs of swimming pleopods, then two for jumping, and then the uropod pair. Thus
in respect of limb differentiation, the amphipods represent the most extreme development of a peracarid stock from the *caridoid facies*. In the
group as a whole, both chelae and subchelae are developed and used
in some forms for climbing on seaweeds, burrowing in the substrate, or
constructing dwelling tubes, as well as in handling food. All carry their
eggs and developing young in a brood-pouch formed of oöstegites, and
there is direct development, the young resembling miniature adults.
The characteristic gammarids include species living in the marine littoral, in estuaries, and in fresh waters—there are even a few tropical
land-dwelling forms. A few genera show some modification of the
characteristic body shape. *Caprella,* the predaceous ghost shrimp living
on littoral weeds and hydroids, has very elongate limbs and thorax,
with a reduced abdomen (Figure 6·5). *Cyamus* is a characteristically
flattened ectoparasite on whales. There are no greatly modified parasites in the order Amphipoda.

To recapitulate, the most highly successful forms in the division
Peracarida are the isopods and amphipods, which are the two groups
most completely—though independently—modified from the primitive
malacostracan pattern.

Archetypic Eucarida

The major division Eucarida encompasses the most highly organized
crustaceans and the largest living arthropods. They are highly successful animals with large numbers, both of species and of individuals.
There are usually several distinct larval stages and metamorphoses in
the life history, and there is a cephalothoracic carapace fused to all 13
of the segments of head and thorax. There is no brood-pouch, and no
lacinia mobilis on the mandible. Once again, within the division we
find a series of progressively more specialized forms ranging from close
to the *caridoid facies* to the true crabs. Systematically, the group is divided into two orders (see Table 5·1), the archetypic euphausiids and
the "advanced" decapod crustaceans.

There are only about a hundred species in the order Euphausiacea, but many of them are of great importance in the marine plankton. They are shrimplike animals of moderate size (2–5 centimeters long), which, except for the eucarid carapace and the larval development, look very like mysids. All the thoracic limbs are similar and biramous with none greatly modified as maxillipeds. They have a single series of small gills at the bases of these legs, and this may involve a respiratory limit upon the size to which euphausiids can develop (see below). The young hatch as a nauplius and then pass through zoealike larva and a mysidlike form. Euphausiids are certainly closer to the *caridoid facies* than are any decapod crustaceans. The feeding mechanism in typical genera such as *Meganyctiphanes* and *Euphausia* is principally by means of a conical basket which is formed by the inwardly developed setae on the endopodites of the first six thoracic limbs. These are held somewhat stiffly as the animals swim along; water is scooped into the conical basket and filtered through the setae. In many forms, including *Euphausia superba,* the bristles are arranged so as to form two or three distinct mesh sizes in different parts of the filtering walls of the basket. In *Meganyctiphanes,* there is a maxillary filter, like that of mysids, which is used in addition to the thoracic basket.

All euphausiids are marine, and strictly animals of the deep-sea plankton. They can occur in relatively dense swarms many miles in horizontal extent, as does *Euphausia superba* in antarctic waters, and they have long been known to whalemen as "krill," the principal food of whalebone whales in certain seas and at certain seasons.

Eucarid Success

The order Decapoda comprises more than eight thousand five hundred species of which four thousand five hundred are crabs (suborder Brachyura). It includes all the familiar forms: shrimps, prawns, crayfish, lobsters, and crabs, and all the biggest crustaceans. There are always three pairs of maxillipeds, and therefore five pairs of remaining thoracic legs (thus the group name)—usually as one pair of chelae and four pairs of walking legs (Figure 6·1). There can be up to three complicated series of gills—podobranchiae, arthrobranchiae, and pleurobranchiae—the details differing in various decapod groups. The exopodite of the second maxilla is modified as a bailing paddle whose continuous action pumps water through the gill series. They have the most highly organized central nervous system found in crustaceans, and can show complex behavior patterns. Most have a series of larvae, though usually the nauplius stage is suppressed and the eggs hatch as zoea (Figure 5·7). The chitinous lining of the stomodaeum is developed as a masticatory apparatus, and all the rest of the exoskeleton is increasingly impregnated with calcium carbonate. There is considerable controversy about the subclassification of this huge and successful group. The four suborders set out in Table 5·1, and used here, are a compromise arrangement and probably unsatisfactory as a complete reflection of phylogeny. In general, the macrurous groups are closer to the euphausiids and more like the *caridoid facies*.

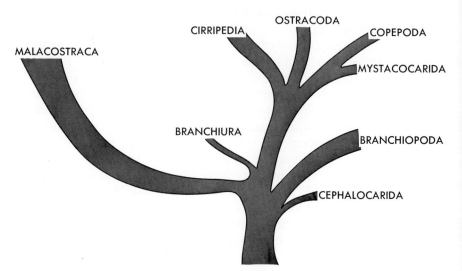

Figure 6·6. A compromise phylogeny of the Crustacea, partly modified from Howard L. Sanders. For discussion, see text.

The first suborder is the Macrura-Natantia, where the abdomen is relatively large and usually extended and there are well-developed pleopods used for swimming. The carapace (and whole body) are often compressed laterally, and the abdominal segments are approximately equal. This group includes all the familiar prawns and shrimps of such genera as *Palaemon, Penaeus, Crangon,* and *Palaemonetes* (Figure 6·2B).

The Macrura-Reptantia are the crayfish and lobsters, where the abdomen is large and usually extended but can be flexed under the cephalothorax, the walking limbs are well-developed and the pleopods are reduced and not suitable for swimming. The first abdominal segment is usually smaller than those posterior to it. Genera such as *Astacus* and *Cambarus* in fresh waters, and *Homarus, Palinurus,* and *Nephrops* in the sea are most extensively eaten by man. Lobsters, such as *Homarus americanus,* can reach weights in excess of 50 pounds, and are thus probably the most massive animals built on the arthropod ground plan. Obviously, such forms as *Homarus* do not have a molt stage (C4) which is terminal.

In the third suborder, the Anomura, the abdomen is always small but somewhat variable in shape. The tail fan is usually less well-developed than in the macrurous groups and the abdomen is held flexed or is markedly asymmetrical. The group includes the hermit-crabs, such as *Eupagurus* and *Pagurus;* the squat-lobsters, such as *Galathea;* and the mole-crabs, such as *Emerita.* There are also a few forms clearly derived from hermit-crabs, but now free-living, which retain a markedly asymmetric abdomen. These include *Lithodes,* one of the stone-crabs (and the related king-crabs of commerce), and *Birgus,* the terrestrial coconut crab.

Finally, the Brachyura, or true crabs, have the abdomen reduced and carried permanently flexed below the thorax, the pleopods being greatly reduced. The carapace is massive and usually globose, or flat and laterally expanded (Figure 4·5). *Cancer, Carcinus,* and *Callinectes* are typical "crab-shaped" crabs; *Maia, Hyas, Libinia,* and *Macrocheira* are spider-crabs with pear-shaped bodies and long legs; and *Uca* and *Sesarma* are fiddler-crabs where one claw of the male is greatly enlarged. Specimens of the Japanese spider-crab, *Macrocheira,* are displayed in many museums: the fact that their spread legs can extend to 9 feet across makes them the world's largest living arthropods.

As already noted, there can be a full series of larval stages in the lower decapods, the ontogeny from nauplius via zoea (Figure 5·7) involving addition of segments posteriorly (see Chapter 2) and the replacement of foliaceous limbs by the stenopodous ones of the adult. In higher forms, the earlier larval stages may be suppressed. In cases where the larger eggs are carried by the females, there is of course no brood-pouch, and they are cemented to setae on the pleopods, giving the characteristic berried appearance.

Various phylogenies of the class Crustacea have been set out. The version presented in Figure 6·6 is a compromise one, but takes into account the evidence resulting from recent investigations on the cephalocarids and other primitive forms. In relation to one topic emphasized in the preceding chapters, we can attempt to set out the possible evolution of feeding mechanisms within the class, although this is not entirely phyletic. This is done in Figure 6·7. From primitive filter-feeders with uniform phyllopodous appendages can be derived the major successful groups of the "lower" Crustacea—the barnacles, copepods, and cladocerans—along with the parasites evolved from them and also (via the *caridoid facies*) the two major evolutionary lines of the malacostracan crustaceans. There is one additional aspect which is not brought out in the figure, and that is the markedly larger size of the successful decapod crustaceans (at the bottom right-hand part of Figure 6·7). The tendency of the eucarid line to move away from the

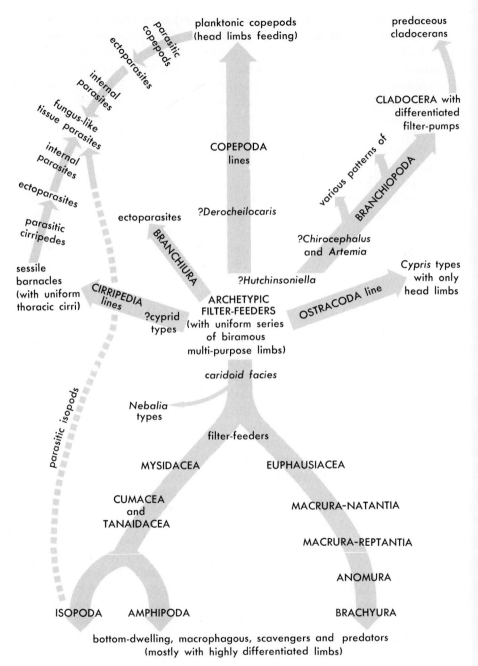

Figure 6·7. The possible evolution of feeding mechanisms within the Crustacea. This does not necessarily reflect phylogeny throughout. The feeding mechanisms characteristic of each "line" are discussed at appropriate places in Chapters 5 and 6.

caridoid facies by becoming macrophagous, bottom-crawling animals has—along with the development of more extensive gills and an efficient system for creating a respiratory current, and the greatly increased impregnation of the exoskeleton with calcium carbonate—allowed the group to evolve an ever-larger body size. Thus we have one of the most successful groups of predatory carnivores in the sea, and the largest living arthropods.

7

Terrestrial Arthropods and Others

THE OTHER TWO MAJOR, successful groups of arthropods are the classes Arachnida and Insecta (numbering more than forty-seven thousand and more than seven hundred thousand species respectively). Unlike the crustaceans, arachnids and insects are primarily land animals, with their whole concert of structures and functions adapted for life in air. In typical winged insects, such physiological processes as aerial respiration, control of water losses, and methods of terrestrial locomotion show efficiencies (in energetic terms) unsurpassed by those of any other organisms. The only other animal groups which approach them in their degree of adaptation for life on land are the amniote vertebrates: reptiles, birds, and mammals. Although the most probable ancestors of presently living arachnids were aquatic arthropods (see p. 123), among both arachnids and insects, the few nonterrestrial forms (almost all found in fresh waters) show *secondarily* acquired adaptations, both structural and physiological, for aquatic life.

The contributions of both the physico-chemical and the mechanical properties of the integument to the success of land arthropods has already been discussed (see Chapter 4). The epicuticle is responsible for the physiological separation of the tissues of a land arthropod from its aerial environment. The mechanical properties of the procuticle and of its articulations are basic to the precise and powerful levering actions which make locomotion on land—and flight—possible. There remain certain other features of arthropod physiology which involve specifically terrestrial adaptations. There are several good textbooks of entomology, largely concerned with describing systematically the enormous diversity of species which are built on the rather stereotyped insect plan. Perhaps as a consequence, many general accounts of invertebrate biology omit all mention of insects. This book is largely concerned with a comparative survey of the functional homologies in invertebrates, and therefore a compromise treatment is adopted, discussing not insect diversity, but those aspects of physiology which are peculiar to land arthropods. Since the physiology of arachnids has, in general, received less detailed investigation, the greater part of this chapter concerns insects.

Insect Respiration

Mechanistically, respiratory organs in land animals are constructed to allow appropriate gas exchanges with the environment (uptake of oxygen, release of carbon dioxide), while minimizing the concurrent water losses. If one excludes minute animals, only three patterns of air-breathing apparatus have been evolved in all terrestrial animals. Lungs have been independently evolved in pulmonate snails and in the higher vertebrates. Tracheal systems, involving air-tubes leading into the tissues, are the only other successful respiratory organs and are found in the majority of land arthropods, including all insects. A third pattern of air-breathing apparatus—sets of lung-books—is found in some arachnids.

The structure and functioning of lung-books can best be studied in scorpions where there are four pairs. Each is an invaginated pit, opening through a narrow slit to the exterior, and containing a large number of leaflike lamellae like the pages of a book. Respiratory exchange depends on diffusion, but the arrangement of the respiratory surfaces—which are the "pages"—results in a relatively small loss of water to the atmosphere. Lung-books have evolved from the gill-books, similarly structured but externally exposed, found in certain primitive aquatic arthropods like *Limulus*. The occurrence of lung-books in other arachnids will be discussed below.

Systems of tracheae are the successful respiratory organs of the majority of successful land arthropods. It is important to realize that tracheal systems represent the best functional adaptation of animals with an arthropod exoskeleton to terrestrial respiration. They have no narrow phyletic significance. Tracheae have been evolved independently by insects, isopod crustaceans, and spiders, by probably at least three other arthropodan stocks, and by onychophorans. The detailed structure and functioning of tracheal systems has been most thoroughly investigated in insects. Tracheae are essentially air-tubes branching from trunks into the tissues of the arthropod, and involving extensions inward of the exoskeleton. Their cuticular lining is thickened in ridges forming a series of annular rings, or a continuous spiral. These serve mechanically to hold the air-tubes open even if pressure within them is somewhat reduced. As extensions of the integument, the entire lining of the tracheal system has to be molted at each ecdysis. A few of the larger tracheate arthropods, including some insects, have muscular pumping of air resulting in rhythmic ventilation of the major tracheal trunks. Many, including the majority of insects, do not pump but can control the rates of diffusion of gases through the tracheal system by adjustment of the external openings or spiracles. Like ventilation rates in higher vertebrates, the control of spiracle opening in some insects is based on detection of increased concentrations of carbon dioxide. Closure of spiracles is used by many arthropods to cut down losses of water vapor while resting, that is, while oxygen requirements are low.

The tracheal tubes branch into ever-finer subdivisions to supply all parts of the body and end in fine air capillaries, termed tracheoles, which are not lined with cuticle. These minute tubes pass intimately among the tissues, usually ending blindly *within* cells including muscle fibers. The walls of the tracheoles are, of course, not molted. As V. B. Wigglesworth first discovered, the extent of gas-filled tracheole can vary with functional demands. In a resting insect muscle, much of the tracheole is fluid-filled. In active muscles and, significantly, in fatigued muscles, there is fluid only at the tips of tracheoles and air can be seen to extend right into the muscle fibers. The physical mechanism of this was elucidated by Wigglesworth, and involves the osmotic pressure in the tissues surrounding the tracheoles acting in opposition to the capillarity forces within the fine tubes. When there has been increased muscle contraction, the tracheole is surrounded by hypertonic fluids (metabolites such as lactic acid having accumulated locally), and fluid is withdrawn from the tracheole into the tissues, air moving in to replace it (Figure 7·1A2). In contrast, in a resting piece of muscle tissue, there are hypotonic fluids round the tracheole and fluids diffuse back into the tracheole lumen to fill most of the terminal tubules (Figure 7·1A1). In both the fluid-filled tracheole and the gas-filled tracheole, oxygen diffuses to an extent determined by concentration gradients, but the rate of diffusion in gas-filled tracheoles will be much higher. Thus the extension of the gas-filled part of the tracheole deeper into the tissues in the active state will improve the supply of oxygen to the cells. As several physiologists have commented, the process is analogous to the dilation of blood capillaries in active vertebrate muscles.

The physiology and ecology of those insects and insect larval stages which live in fresh waters have many interesting aspects. The terrestrial tracheal system is readapted for aquatic life in many ways, as in the similarly evolved aquatic pulmonate snails (see BLI, pp. 132–134). Some aquatic insects are merely divers, surfacing at regular intervals to breathe air. Others use an exposed gas bubble as a "physical gill" to extract oxygen from the water by diffusion and pass it thence into the tracheal system. In several insects, this process has evolved into plastron respiration, in which a thin film of gas is carried between minute hydrofuge hairs to form a physical gill which does not require renewal at intervals. Other insects have become fully aquatic by developing tracheal gills in which fine ramifying tracheae lie immediately below an extremely thin cuticle.

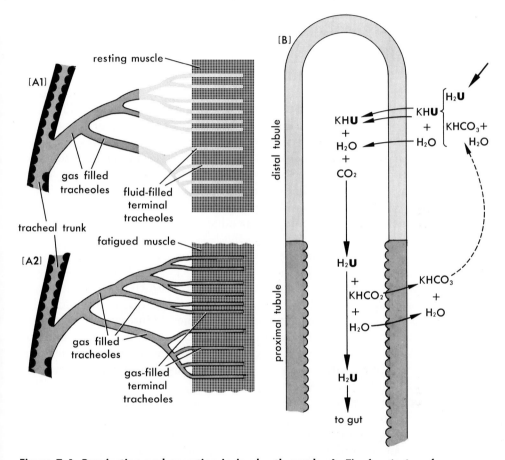

Figure 7·1. Respiration and excretion in land arthropods. A: The functioning of tracheoles: in **A1** fluids fill most of the terminal branches in resting muscle tissue, where there are hypotonic fluids surrounding the tracheoles; in **A2** the terminal tracheoles are gas-filled for more efficient respiratory exchange in active or fatigued muscle tissue, where the accumulation of metabolites has rendered the fluids surrounding the tracheoles hypertonic, bringing about withdrawal of fluid from each tracheole. **B:** The functioning of the excretory organs of insects, the Malpighian tubules. The processes within the two histologically distinct regions of a single tubule are illustrated. They lead to the expulsion of the insoluble crystals of uric acid into the gut, following the transport of the relatively soluble potassium acid urate into the lumen of the distal part of the tubule. For further discussion, see text. [Modified from various figures by V. B. Wigglesworth.]

Two final points about tracheal systems concern the evolution of arthropods. Since they depend on gas diffusion, active arthropods using tracheal respiratory systems are probably limited in size. It is almost certainly this, rather than the mechanics of the exoskeleton, which limits the maximum size of insects. The other matter which must be recapitulated is that, in possibly seven or eight lines of arthropods, tracheal systems have been independently evolved in the course of colonization of land.

Insect Excretion

In several groups of land arthropods, there are organs which correspond to the segmentally arranged coelomoducts of annelids. The coxal glands found in several groups of arachnids and other arthropods are homologous with the excretory organs in crustaceans. However, the important functional excretory organs of the best-adapted land arthropods are not such segmental structures but diverticula of the gut, called Malpighian tubules. These are the organs of nitrogenous excretion in insects, and are also found in millipedes, centipedes, and several groups of arachnids. Once again, independent evolution of these structures is most likely.

In such land arthropods with Malpighian tubules, nitrogenous excreta are mainly in the form of uric acid crystals. A solid or semisolid urine of this composition is, of course, also characteristic of the water-conserving vertebrates: reptiles and birds. Once again, the functioning of Malpighian tubules has been most studied in insects, and best elucidated by V. B. Wigglesworth. There are two distinct histological regions in each tubule: a proximal portion where the lining cells are differentiated with a brush border and a distal region where they are less differentiated and are said to have a honey-comb border. The processes leading to expulsion of uric acid crystals into the gut, and thence to the exterior with the faeces, will be best understood with reference to Figure 7·1B. Uric acid can be combined with potassium bicarbonate and water to form the relatively soluble potassium acid urate, which is actively transported from the hemocoel into the lumen of the tubule in its distal section. The contents here are neutral or faintly alkaline, while in the proximal part they are acid. In the lumen of the proximal part, potassium bicarbonate and water are reabsorbed, resulting in the precipitation of uric acid crystals.

Combined with this mechanism, there is usually resorption of water in the gut, and in many insects the blind distal ends of the Malpighian tubules are closely applied to the water-resorbing section of the hindgut. In other cases, the Malpighian tubules lie within blood sinuses. Once again, each aspect of the integrated concert of functions in a well-adapted terrestrial animal must involve conservation of water.

The efficiency of the interacting lever-systems of the arthropodan machine—the jointed limbs—is nowhere so manifest as in locomotion on land. The limbs of insects, arachnids, and the myriapodous groups are all—to use crustacean terminology—stenopodous and uniramous. These elongate systems of rigid, tubular sections and articular membranes have their condyles (and thus also their axes of articulation; see Chapter 4) arranged in specifically precise orientation. The structural arrangements permit certain mechanical efficiencies.

The slender limb with its pointed tip allows the propulsive force to be applied through a single point on the substrate (through the *point d'appui*) and, in the stepping action of a series of limbs, allows each successive limb to take over almost the same *point d'appui* in turn. A limb length greater than the diameter of the trunk, combined with the specific condylar axes (see Figure 7·2B), not only allows the body to be carried "high"—clear of all contact with the substrate in most land arthropods—and slung between the limbs in a stable fashion, but also allows for controlled changes in the limb's effective length.

This last factor is of paramount importance in mechanically efficient stepping, allowing the distance between the mid-line of the arthropod and the *point d'appui* of the limb (M–P in Figure 7·2A, B) to remain constant while the trunk moves forward during each step. (This ensures that no propulsive energy is wasted in lateral swayings of the trunk.) The actual propulsive force can be generated either by the contraction of muscles largely extrinsic to the limb itself, acting across its proximal articulations (the case in the majority, and probably also in the most primitive terrestrial limbs), or by contraction of intrinsic muscles acting across more distal joints. In both cases, the intrinsic musculature—flexors and extensors—is responsible for the continuous changing of the angles between podomeres during the propulsive backstroke of the step. Such continuous adjustment in the effective length of the limb allows the M–P distance to remain the same for the duration of the backstroke, so that effective propulsion is maintained in the direction of the mid-line. To put it another way, the distance between the limb tip and its basal articulation on the trunk *must be shortest* at the mid-backstroke. In each limb, the propulsive backstroke is followed by a recovery movement forward with the distal parts raised clear above the substrate, which recovery stroke ends with the limb tip at a new *point d'appui*. In almost all terrestrial arthropods, the limb is overall less flexed (i.e., has more extensor muscles contracted) during the recovery stroke and so the limb tip swings forward in an arc, much further from the trunk (M–R on Figure 7·2A, B) than the line of *points d'appui* (the "footprint" line). Again in a generalization, the muscles, both intrinsic and extrinsic, which are contracted during the "power"

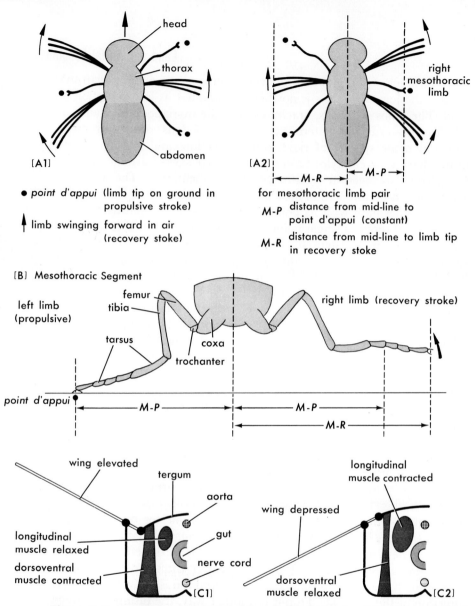

[A1]

head

thorax

abdomen

• *point d'appui* (limb tip on ground in
propulsive stroke)

↑ limb swinging forward in air
(recovery stoke)

[A2]

right
mesothoracic
limb

M-R ——— M-P

for mesothoracic limb pair

M-P distance from mid-line to
point d'appui (constant)

M-R distance from mid-line to limb tip
in recovery stoke

[B] Mesothoracic Segment

left limb
(propulsive)

femur
tibia

tarsus

coxa

trochanter

right limb (recovery stroke)

point d'appui

M-P ——— M-P

M-R

wing elevated

tergum

aorta

longitudinal
muscle relaxed

dorsoventral
muscle contracted

gut

nerve cord

[C1]

longitudinal
muscle contracted

wing depressed

dorsoventral
muscle relaxed

[C2]

Figure 7·2. Locomotion in insects. A1 and **A2:** Successive "steps" in a stylized
insect (basically dipteran) which is walking slowly. The leg pairs are in opposite phase
and so the insect is always stably supported as a tripod upon three *points d'appui.*
All limb tips are further from the mid-line of the body during recovery strokes (M–R),
than while they are exerting propulsive force (M–P), and the mesothoracic limbs
work on wider spaced lines of *points d'appui* (i.e., leave wider footprint lines). **B:**
Left and right mesothoracic limbs in a stylized insect (basically dictyopteran) diagrammed during a "step" in which the left limb is propulsive. **C1** and **C2:** Successive
positions of the indirect flight muscles in a stylized insect (basically hemipteran),
showing the dorsoventral muscles of the thorax as antagonists of the longitudinal
muscles. This is an oversimplification of the mechanics of wing movements in aphids:
see text for further discussion.

backstroke of the limb, will anatomically be more massive than their antagonists. Likewise, the muscles which are contracted to bring about the forward swing of the recovery stroke will be slighter and less powerful.

Essentially this mechanical pattern is utilized by land arthropods in a wide variety of locomotory activities: slow crawling, fast running, digging, jumping, climbing silk threads or vertical surfaces, and even running while hanging upside down from ceilings. This variety is achieved not so much by change in the functional pattern of power and recovery strokes (except for jumping) as by anatomical differences in the proportions of the podomeres of limbs, in the distances between articulations, and in the pattern of extrinsic limb muscles. Mechanically, what differ are the proportions of each lever arrangement around its fulcrum.

The locomotory mechanisms of land arthropods probably evolved from those of an annelidlike ancestor with large parapodia used in crawling (as in present-day *Nereis* involving power and recovery strokes in a metachronal rhythm with the opposite sides of each segment exactly out of phase—see BLI, pp. 110–111). An intermediate stage would be represented by the undifferentiated locomotion of presently living walking-worms (Onychophora—see p. 125), with soft worm-like bodies carried upon a series of paired limbs. The locomotory sequences in walking-worms like *Peripatus* can be related to those in the myriapodous land arthropods. Finally, reduction in limb numbers brought about the efficient eight- and six-legged patterns of arachnids and the six-legged pattern of all insects. This sequence—errant-polychaetes to walking-worms to myriapods to insects—certainly represents the progressive specialization in land locomotion which must have occurred—though, of course, these four groups of living animals cannot represent the actual evolutionary lineage, and may not reflect phylogeny at all closely. Present understanding of the probable course of locomotory specialization in land arthropods is almost entirely based on a beautiful and extensive series of studies on many types of myriapods and on *Peripatus* by Dr. Sidnie M. Manton, which studies combine, in the best traditions of functional morphology, superbly detailed accounts of the microanatomy of muscles and of exoskeleton with a detailed mechanical analysis of the locomotory sequences in the healthy living organism. Her papers, mostly published in the British *Journal of the Linnean Society of London* (*Zoology*), should be consulted by interested students, not only as a detailed account of structure and function in the locomotion of myriapods which cannot even be adequately summarized in a book of this size, but as an example of how relatively simple observational work (using analysis of flash photographs and of motion picture film, and measurement of tracks made on smoked

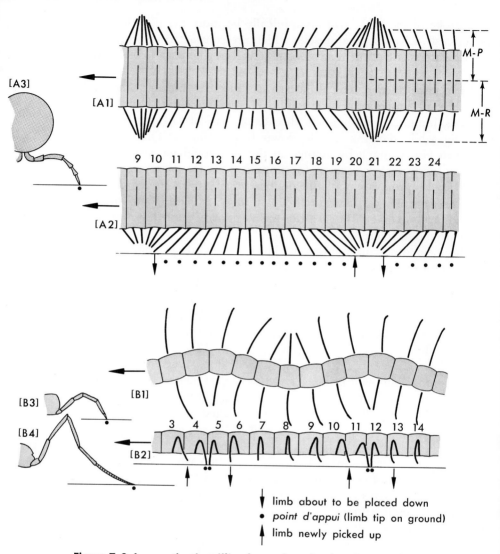

9 10 11 12 13 14 15 16 17 18 19 20 21 22 23 24

3 4 5 6 7 8 9 10 11 12 13 14

↓ limb about to be placed down
• *point d'appui* (limb tip on ground)
↑ limb newly picked up

Figure 7·3. Locomotion in millipedes and centipedes. A1: Dorsal view of a typical millipede (such as *Spirostreptus* or *Gymnostreptus*) in motion, showing sixteen of the short, wide diplosegments with thirty-two pairs of limbs, left and right sets being exactly in phase with each other. **A2:** Lateral view of the same millipede, showing that the majority of the limb tips are on the ground at once giving the *longer* strokes of propulsion characteristically "low-geared like bulldozers." **A3:** The millipede limb, relatively short and attached nearly mid-ventrally to the short, wide, cylindrical diplosegment. **B1:** Dorsal view of a typical centipede (such as *Scolopendra* or *Cryptops*) in motion, showing twelve of the long, narrow segments each with a pair of limbs in opposite phase and the undulations of the body which occur at speed and are mechanically inefficient. **B2:** Lateral view of the same centipede, showing the larger phase difference between segments with less than a third of the limb tips on

paper) can lead to an understanding of the mechanics of complex muscle systems and to hypotheses on their evolution.

As a relatively simple example of the mechanical possibilities of "many-legged" systems in land arthropods, the differences between a typical millipede (class Diplopoda) and a typical centipede (class Chilopoda) can be considered (see Figure 7·3). Millipedes are somewhat slow-moving, requiring relatively powerful propulsion as they move continuously through soil and decaying vegetable matter and ingest some of it. Centipedes are fast-moving, free-running animals, which are predaceous carnivores searching for and running down their prey. The type of gait shown by the former has been analyzed by Manton and can be termed "low-geared" while the gait of centipedes, as first elucidated by E. Ray Lankester early this century and reinvestigated by Manton, is "high-geared." For equivalent motor size (assessed as volume of locomotory muscles), millipedes are arranged with gearing like bulldozers, centipedes like racing cars. Figure 7·3 illustrates some of the ways these differences of gait are achieved in a millipede like *Spirostreptus* and a centipede like *Scolopendra* (or *Cryptops*). The most significant functional difference is that the relative durations of the propulsive backstroke and the recovery stroke are 7 to 3 for *Spirostreptus,* and less than 3 to 7 for *Cryptops:* longer propulsive strokes for power, shorter propulsive strokes for speed. This correlates with differences in the metachronal succession of limb movements (see Figure 7·3), the phase difference between successive legs being very small for the millipede and relatively large for the centipede. Structurally, the millipedes have many short, wide segments and many relatively short legs, while centipedes have fewer long, narrow segments with (in *Cryptops*) somewhat longer legs (and, in other centipedes, much longer legs). One result is that at any one time in the millipede, the majority of the limb tips are on the ground, while in the centipede, less than a third of the limbs are simultaneously supporting (and propelling) the trunk. Thus, while it is possible for the millipede to have its pairs of legs in phase with each other, in the centipede the gait almost demands that the paired legs be in opposite phase: to ensure stability and support of the trunk without sagging. Thus we see that the powerful bottom gear of millipedes involves long backstrokes with many legs pushing simultaneously (the metachronal waves involving

the ground at once, giving the short swift strokes of propulsion characteristically "high-geared like racing cars." **B3:** The centipede limb (in *Scolopendra*), relatively long and attached laterally to the longer, thinner segment. **B4:** The much longer limb of the common house centipede, *Scutigera*. For further discussion, see text. [Partly adapted from E. Ray Lankester in *Quart. J. Micros. Sci.*, 47:523–582, 1904 and from S. M. Manton in *J. Linn. Soc. (Zool.)*,42:93–166, 1952, and 299–368, 1954.]

large numbers of pairs of legs). In contrast, the faster gait of centipedes involves few legs propelling at once by backstrokes of short duration. They show some sacrifice of stability to speed, and a tendency to reintroduce the mechanical inefficiency of lateral undulations of the trunk. Swaying of the trunk segments to each side alternately involves energy waste, and results from the propulsive limbs being widely separated and alternately placed.

Turning to the more successful land arthropods, spiders and insects, we have a significantly higher degree of tagmatization, with a smaller number of limbs, each relatively longer than those in myriapods. Significantly, the articulations of the three or four pairs of legs with the trunk are always close together near the center of gravity of the body. For example, in the insects the central tagma, the thorax, consists of only 3 segments—prothorax, mesothorax, and metathorax, usually fused but each bearing a pair of limbs. Thus, mechanically, in spiders and insects the propulsive force of the few long legs is applied to the body near a single center, avoiding any undulatory tendency. In general, insects stand or walk with the leg pairs moved in opposite phase (see Figure 7·2A), and thus these hexapodous organisms are stably supported as tripods, the left mesothoracic leg being on the ground with the right prothoracic and metathoracic limbs. A few spiders move habitually on six legs like insects, while in others, using eight legs, there is a phase difference of a quarter of a cycle between one leg of each pair and its partner.

For obvious reasons of both static posture and locomotion, in arthropods with a reduced number of longer legs attached close to a single center, the legs must fan out anteriorly and posteriorly. This is as true of decapod crustaceans as it is of insects and spiders. Locomotor efficiency is increased by separating the *points d'appui* of the limbs, but this also leads to a functional differentiation not often discussed. Taking a characteristic insect, for example, and ignoring for the moment the middle (mesothoracic) pair of legs, the forelegs and the hindlegs execute their propulsive backstrokes by strikingly different muscle contractions. Apart from the extrinsic limb muscles (which are also differentiated), most of the intrinsic flexors of a prothoracic leg will be contracting during the propulsive backstroke (in other words, a front leg ends its power stroke and begins its recovery stroke more flexed than at the beginning of its step). In contrast, contraction of the intrinsic extensors is involved in the propulsive stroke of a metathoracic leg (that is, the leg ends its power stroke more extended). If the earlier generalizations are accepted, then the attentive reader can deduce the size differences to be found in the muscles of the fore- and hindlimbs of insects.

The thorax of most insects also bears wings: archetypically two pairs on the meso- and metathoracic segments, but reduced to a single functional pair in several distinct orders. The mechanics of insect flight are still in part controversial, and there are several variant patterns. Only a partial synopsis can be attempted in a book this size. The simplest case, mechanically, can be illustrated in a stylized model aphid (see Figure 7·2C), in which the articulation of the wing is greatly simplified. In this, the flapping of the wing results from the antagonistic contractions of the indirect flight muscles which can deform the mesothoracic and metathoracic segments, mainly by moving the fused terga (dorsal roof plates) of the thorax. As shown in the aphid mesothorax (Figure 7·2C) contraction of the dorsoventral muscles elevates the wings by pulling down the tergum and at the same time slightly elongates and laterally extends the thoracic box, thus stretching the longitudinal muscles. Then contraction of the longitudinal muscles shortens the thoracic box, the tergum moves upward, depressing the wings and stretching the now-relaxed dorsoventral muscles. As these two sets of indirect flight muscles work in rhythmic alternation, the so-called direct flight muscles (attached directly to the wing bases at their articulation) alter the "angle of attack" of the wings so that qualitatively they can be acting as lifting aerofoils throughout the cycle. To achieve this lift by having appropriate small angles of attack during both downstroke and upstroke, the direct flight muscles twist the wing so that its leading edge is twisted downward during the downstroke and upward during the upstroke. This cycle of change in the angle of attack seems to be true for all medium and larger insects in flapping flight forward at moderate rates of wing beat, and is accompanied in many by changes in wing camber also brought about by the indirect flight muscles. (It is obviously different in hovering flight, at faster beat frequencies, and in very small insects.) Insect flapping flight has received detailed experimental and kinematic studies only in the desert locust (by T. Weis-Fogh and Martin Jensen), and therein the principles of maintained aerofoil action outlined above hold true (although the muscle systems are different from the aphid), and analysis can follow the principles of simple propeller theory. Crudely summarized, during the downstroke all of each wing's length contributes to lift, and the outer parts also an excess of thrust, while during the upstroke, much of the wing's length still contributes to lift but there is no thrust. Thus, under many circumstances, much less power is required for the upstroke than for the downstroke, and in certain cases the wing is actually raised by the wind (the resultant of the forward motion), and so there can be a *negative* power

111

requirement for the upstroke. Such is the case in the desert locust where calculated figures of the power required for aerodynamic work in the downstroke and in the upstroke were found by Jensen and Weis-Fogh to be in reasonably close accordance with experimental measurements of work done. However, this basic pattern of aerodynamics is modified in many insects.

One difference of some evolutionary significance is that in the more primitive groups of flying insects such as the dragonflies (Odonata), contractions of the direct flight muscles supply the forces for the down-stroke and upstroke. Depressor and elevator muscles for both pairs of wings can be distinguished anatomically. The neuromuscular arrangements of such muscles are also simple: each contraction resulting from a single motor nerve impulse, the wing-beat rate being about 20–28 per second, and being termed synchronous. In contrast, the faster small insects belonging to groups usually thought less primitive, such as Hymenoptera and Diptera, neuromuscular function is asynchronous, each nerve impulse setting up an active state in the muscle, during which a variable number of contractions (perhaps 40) can occur. The muscles exhibiting this myogenic mechanism are termed fibrillar. Frequencies of wing beat can thus be high: 250 per second for hive-bees; 200 per second for *Musca*, the housefly; and even about 1,000 per second in some ceratopogonid midges.

Perhaps surprisingly, the flight muscles of insects (even including these fibrillar muscles) are neither unusually strong nor do they show unusually rapid shortening when compared to those of other animals including vertebrates. However, since power is measured as work per unit time, the high frequency of twitches in the synchronous muscles and even more the high frequency produced myogenically in the fibrillar muscles result in unusually high values for power output. These are reflected in the oxygen consumption and in the very high rates of utilization of carbohydrates and of fats as metabolic energy sources in flying insects. The flight muscles all work aerobically, and there is never any fatigue (or accumulation of intermediary metabolic products) in them as can occur in active locomotory muscles of other animals. Adaptively, this aspect of insect muscle physiology reflects the need for flapping flight to be sustained. A flying insect could not stop in mid-flight to allow oxidative recovery from an accumulation of lactic acid!

Finally, in many insects—perhaps most—power is saved in flight by various mechanisms of elastic recoil. E. G. Boettiger has provided a detailed analysis of the wing-hinge system in the dipterous fly, *Sarcophaga*, where the skeletal arrangement is such that when the wing is in any intermediate position between greatest elevation and greatest depression, the curved surfaces of the skeletal elements are strained, involving storage of energy. Their recoil can "click" the wing to the up position, and this "click" or "snap" mechanism like that of a child's metal cricket ensures a magnification of each muscle contraction and a rapid movement to each extreme position in turn. Thus the minimum change of length required of the fibrillar flight muscles (with an amplification of about 500 times at the wing tips) allows them to operate nearly isometrically. A more obvious example of intrinsic elasticity concerns the cuticle of the hinge in some insects (first studied in locusts by Weis-Fogh). This hinge material is largely a peculiar rubberlike protein, called resilin, with cross-link bonding of a type unique in structural proteins. Mechanically, this intrinsically elastic hinge allows the energy of the "negative work" of the upstroke already mentioned to be stored, and to contrbute to the following downstroke of the wing. Finally, the entire flight muscles have been shown to possess a strong passive-elastic component themselves. All such mechanisms of elastic recoil enable the insects possessing them to minimize the dimensions of their flight muscles. In summary, the locomotory mechanisms of land arthropods include the most sophisticated patterns found in animals, and these are all functionally based on the unique properties of the arthropod exoskeleton.

Other Insect Functions

As has been seen, terrestrial adaptation and the nature of the exoskeleton determine the patterns of respiration, excretion, growth, and locomotion in insects. A brief consideration of the process of nutrition, circulation, and reproduction seems appropriate here.

The insect alimentary canal is a simple tube, though often with some convolutions, running from the mouth to the anus. Structurally and functionally, it is divisible into a cuticle-lined foregut, an endodermal midgut, and a hindgut—again with a cuticular lining. In the majority of insects, valvular arrangements separate the three regions, and, if diverticula occur other than the Malpighian tubules, they form gastric caeca attached to the anterior end of the midgut, or a blind storage crop attached to the foregut (as in many Diptera, butterflies, and moths).

The form of the foregut varies with the diet of insects. Various forms of mouth parts are, of course, associated with each of the major orders of insects, but functionally they fall into three groups. Most primitive are chewing and biting mouth parts (suitable for predaceous carnivores, foliage-eaters, and other macrophages); while more specialized are those for collecting fluids (products of bacterial decomposition, or angiosperm nectars); and those adapted for piercing and sucking (not only animal blood but also tissue fluids of plants). Functionally, the foregut is primarily a storage organ, with a crop formed as a dilation or a diverticulum, and a capacity appropriate to dietary habits. There

can also be a "gizzardlike" region for internal trituration. Some digestion may take place here, the enzymes originating elsewhere, but there
is no absorption. Functionally, the main site of production of digestive
enzymes, of digestive breakdown, and of absorption, is the midgut.
Being lined, as are its caeca, with endodermal epithelia, there is no
locally secreted epicuticle and procuticle as in the other two gut regions.
A peculiar loose lining—the peritrophic membrane—is secreted in
various ways, is functionally semipermeable, and although extremely
thin, is probably of similar composition (of chitin and protein) to the
deeper layers of the procuticle in· the exoskeleton. It permits passage
of liquids and solutes in both directions but apparently prevents solid
particles of food from coming into contact with the gastric epidermis.
Thus, unusually clearly demonstrated for invertebrates, digestion is
entirely extracellular. The hindgut is again lined with invaginated ectoderm and therefore with an exoskeletonlike cuticle. Its major function
appears to be uptake of water from both the faecal remains from the
midgut and the nitrogenous excreta from the Malpighian tubules. It is
worth noting that while the peritrophic membrane is continuously secreted and renewed (in many cases forming a sheath around the faecal
pellets on their expulsion), the cuticular linings of foregut and hindgut
have to be reformed and shed along with the exoskeleton at every molt.

As in all arthropods, the circulatory system is an "open" one, the
heart being tubular with a series of ostia and the only true blood vessel
being the anterior aorta leading toward the head from the heart. Blood
is released anteriorly and returns through a series of sinuses to the pericardial space and thence through the ostial openings into the heart. In
view of the nature of the respiratory structures, insect blood does not
need to be effectively involved in the transport of oxygen to the tissues.
A large part of the osmotic pressure of insect blood is maintained by
amino acids, and the ionic ratios found in herbivorous insects involve
unusually high potassium to sodium values. This latter fact creates difficulties in applying the usual physiological explanations to some aspects
of neuromuscular function in plant-eating insects. There are large numbers of nucleated blood cells, and about thirty types in six classes have
been described. Most are phagocytic. The ratios of blood-cell types
change with the molt cycle in most insects but the functional significance of many of the differentiated cells remains obscure.

The great majority of insects are dioecious, although parthenogenesis has been described in several groups. The paired testes in the
male usually consist of a series of follicles, within each of which
can be found a series of groups of developing spermatozoa contained in packets. The several stages of sperm maturation are to be
found in a serial arrangement. There are usually storage sacs, termed
seminal vesicles, and in some cases accessory glands. The female system

is similarly arranged for the serial production of gametes. Each ovary consists of a series of fingerlike ovarioles, each of which is an epithelial tube with the oögonial germ cells at the free end and a contained series of eggs in various stages of development. The number of eggs which can be laid at one time obviously corresponds to the number of ovarioles. In many cases each laying is enclosed in an oötheca or other egg capsule or mass secreted by the lower part of the female genital system (oviducts, vestibulum, or vagina) or by various kinds of accessory glands. Hormonal mechanisms are undoubtedly responsible for determining the onset of reproduction in insects, and probably for controlling fecundity (see p. 52). However, the secondary sex characters do not ever seem to be controlled by hormones originating in the insect gonads. The contrast with crustacean endocrine physiology has already been noted.

Arachnid Diversity

The second largest class of the arthropods, the Arachnida, are structurally rather uniform. They all have 21 segments in two tagmata: the prosoma and opisthosoma. The prosomatic segments are completely fused in most forms, and bear the rather stereotyped six pairs of appendages. These are, from the anterior end: the preoral chelicerae— usually prehensile and concerned in food capture—then the pedipalpi which may be tactile sensory organs or relatively enormous chelae like those of crabs. The remaining four pairs of appendages are the walking legs. The opisthosomatic segments show some variation, being clearly separated and divided into two series in primitive forms like scorpions, or completely suppressed by fusion as in the big specialized group, the Acarina, or ticks and mites. Respiration is by lung-books or tracheae, or both, and the excretory organs are coxal glands of coelomoduct type found usually as a single pair in various middle segments of the prosoma or Malpighian tubules (see p. 104). Most modern classifications of arachnids involve ten orders, of which two are of enormous extent— the Araneae, or spiders, and the Acarina, or mites—and five are certainly of minor importance.

An intermediate-sized group, the order Scorpionida, are clearly the most primitive living arachnids, and the oldest (Silurian) known terrestrial arthropods in the fossil record. Scorpions are cryptic and nocturnal animals of the warmer latitudes. Characteristically fearsome animals to man, they have the pedipalps as powerful grasping chelae, and a long flexible abdomen bearing the stinging apparatus with its sharp, hollow, curved barb for venom injection. However, scorpions are predaceous carnivores feeding mainly on insects but also on all other suitably sized small invertebrates. For example, a cockroach

caught by a scorpion of about its own length will be held in the pedipalps while the abdomen is flexed over the back of the body and the
venomous barb stabbed into it. The venom is a mixture of substances
but apparently contains a paralyzing neurotoxin. The form of the
scorpion opisthosoma is very primitive by arachnid standards, and is
made up of a 7-segmented preabdomen bearing a pair of sensory appendages called the pectens and containing the four pairs of lung-books,
and a 5-segmented postabdomen of elongate annular segments with the
last segment bearing both the anus and the stinging apparatus.

Two further orders of intermediate size are the Chelonethida (Pseu-
doscorpionida), or false-scorpions, and the Opiliones or harvestmen.
Chelonethids are minute arachnids living as predaceous carnivores on
small arthropods in the soil. The pedipalps resemble those of scorpions
but have associated poison glands, while the abdomen is relatively wide
and rounded posteriorly. The more familiar long-legged harvestmen,
order Opiliones, number about one thousand eight hundred species.
The legs are always extremely long and slender, and in some tropical
forms reach lengths of 16 centimeters. There is no constriction between
the prosoma and opisthosoma in harvestmen, and the latter shows visible
segmentation. The tracheal system of Opiliones differs from that in
other arthropods and seems to have been independently evolved, and
there are peculiar odoriferous glands borne in the prosoma. Harvestmen are omnivorous animals living mainly in humid habitats. The five
minor orders, none of which shows consistently archetypic features, include: the Solifugae or sunspiders, the Uropygida or whipscorpions,
the Amblypygida or scorpion-spiders, the Palpigradi, and the Ricinulei.
Each of these five groups numbers less than one hundred species.

There remain the two huge and successful orders of the arachnids.
The Araneae, or spiders, number about thirty-four thousand species
and are abundant predaceous carnivores in many types of vegetation
in all parts of the world. The prosoma is joined by a waistlike pedicle
to the usually unsegmented opisthosoma. The chelicerae are of moderate size and bear poison glands. Respiration is by lung-books and associated tracheal tubes. There are usually eight pairs of eyes on the
prosoma, which although they are simple ocelli, are more highly developed than those of other arachnids. There are usually three pairs of
spinnerets on the opisthosoma and these secrete the characteristic silk
threads. The silk is a complex albuminoid protein, which when hardened has great tensile strength and elasticity. All spiders continually lay
a dragline for their own safety. Many spiders build webs to capture
prey, and other uses for silk include lining burrows, enclosing eggs,
encasing sperms at copulation, constructing special copulatory chambers, constructing similar chambers for molting or for going into a
diapause within. Certain chelonethids and mites also spin silk threads.

The other major order, the Acarina, comprises the much more than nine thousand species of mites and ticks. Even apart from the numerous parasitic forms, these are the arachnids of greatest economic or ecological importance to man. Mites can occur in fantastic numbers as destructive pests of crop plants, stored foods, and other natural products. The mite body is completely fused, even the division between prosoma and opisthosoma being imperceptible. Except for the four pairs of legs, the paired appendages are usualy minute and associated with the mouth parts which can be arranged functionally for either biting or piercing and sucking. Larval stages have six pairs of legs. All but the smallest mites have tracheae as respiratory organs. Most mites are minute, and the group includes the smallest adult arthropods where the mature size is less than 0.1 millimeter. One superfamily of the Acarina has secondarily returned to aquatic life in both the sea and fresh waters. Many free-living mites are predaceous carnivores, while others are omnivorous scavengers.

Insect Diversity

Insects are the most numerous land animals and, in terms of number of species (seven hundred thousand), the most numerous class of living animals. As already noted, they are remarkably stereotyped in structure and function, and such differences as do occur between the major groups of insects concern the type of wings, the type of metamorphosis, and the type of mouth parts. On the basis of such characteristics, which *do* have considerable functional and ecological significance, thirty-three orders of insects can be distinguished (see Table 7·1).

These orders are grouped into two very unequal subclasses, Apterygota and Pterygota. The former contains only four minor orders of fundamentally wingless insects. Apart from the primitive apterous condition, they retain paired abdominal appendages and perhaps the most unspecialized mouth parts found in insects. The four orders listed in Table 7·1 include the relatively abundant springtails and bristletails, such as *Lepisma* and *Campodea* (see Figure 7·4). There are about one thousand five hundred species of apterygote insects as against the seven hundred thousand or so of the subclass Pterygota or fundamentally winged insects (among which there are some orders where the wings are *secondarily* reduced or even absent).

The subclass Pterygota can be divided into four divisions, the first of which, the Palaeoptera, comprises the more primitive winged insects in which the wings cannot be folded and are held permanently at right angles to the body. Several extinct orders of insects, including the earliest flying insects, would be included in this division, along with the two living orders, Ephemeroptera (may-flies) and Odonata (dragon-

TABLE 7·1

Synopsis of a Classification of Insects

Phylum ARTHROPODA: Class INSECTA

Subclass **APTERYGOTA**

containing four orders: COLLEMBOLA (Springtails), PROTURA, DIPLURA and THYSANURA (Silverfish and Bristletails).

Subclass **PTERYGOTA**

Division **PALAEOPTERA** containing two orders: EPHEMEROPTERA (May-flies) and ODONATA (Dragonflies).

Division **POLYNEOPTERA** containing nine orders: DICTYOPTERA (Cockroaches and Mantises), ISOPTERA (Termites), PLECOPTERA (Stone-flies), ORTHOPTERA (Grasshoppers and Crickets), DERMAPTERA (Earwigs), and four more minor orders.

Division **PARANEOPTERA** containing seven orders: ANOPLURA (True sucking-lice), HEMIPTERA-HETEROPTERA (True Bugs), HEMIPTERA-HOMOPTERA (Aphids and Leafhoppers), and four others.

Division **OLIGONEOPTERA** containing eleven orders: NEUROPTERA (Lacewings), COLEOPTERA (Beetles), TRICHOPTERA (Caddis-flies), LEPIDOPTERA (Moths and Butterflies), DIPTERA (True Flies), SIPHONAPTERA (Fleas), HYMENOPTERA (Wasps, Ants, and Bees) and four more minor orders.

flies). Both orders have their young stages living in fresh waters as nymphs which undergo only simple metamorphosis as they grow and mature to the adult. Both show rather unspecialized chewing mouth parts and two pairs of primitive membranous wings which cannot be folded. The aquatic nymphs of all may-flies and of some Odonata are structurally "thysanuriform" or "campodeiform" (see Figure 7·4 in which these are compared to adult apterygote insects).

The remaining three divisions are neopterous, being able to fold their wings in various ways. The divisions Polyneoptera and Paraneoptera encompass the winged insects which hatch as nymphs and undergo a series of molts accompanied by only gradual and simple metamorphosis. Their wing buds develop externally in preadult stages, and the groups were included in earlier classifications in the Exopterygota or Hemimetabola. The remaining division, Oligoneoptera (see Table 7·1), encompasses the ten important insect orders where the young hatch as larvae, do not show external wing buds, and undergo a special complex metamorphic molt or molts, associated with an inactive pupal stage preceding the adult emergence. In earlier classifications, the Oligoneoptera were, for these features, termed the Endopterygota, or Holometabola. The division Paraneoptera includes certain forms which are in some ways intermediate between the other two neopterous divisions, and an incipient pupal stage is found in some bugs and other representative families of the division.

The Polyneoptera comprises nine orders, of which the five likely to be familiar to the reader are set out in Table 7·1. The Dictyoptera and Orthoptera, cockroaches and grasshoppers and their allies, are typical of the division with nymphal stages slowly leading to the adult form, with biting and chewing mouth parts, and with relatively great powers of running and jumping but poor powers of flight. They include the largest presently living insects. The stone-flies, order Plecoptera, have aquatic nymphs which are again campodeiform (Figure 7·4).

The division Paraneoptera encompasses seven orders, including the Anoplura, or true sucking-lice, where the wings have been secondarily lost and the adults are parasites on birds and mammals, and the two large hemipterous orders. All three orders have piercing and sucking mouth parts and a gradual metamorphosis toward the adult. Both the

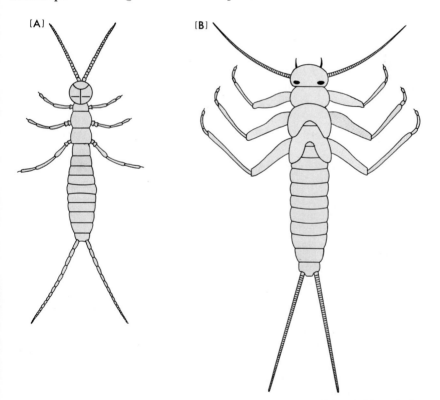

Figure 7·4. A possibly archetypic inspect pattern. A: An adult primitive wingless insect, a "bristle-tail," *Campodea,* about 4 millimeters long. **B:** A late aquatic nymphal stage of the plecopteran "stone-fly," *Perla,* about 3 centimeters long. Note the external wing buds in this preadult stage (the Exopterygota of earlier classifications; see text), and the generally "campodeiform" pattern.

Hemiptera-Heteroptera and the Hemiptera-Homoptera are large diversified groups containing large numbers of species, and some forms, such as aphids, are often extremely abundant as individuals, as a result of fantastic fecundity.

Finally we have the eleven orders of the division Oligoneoptera, or Holometabola, of which seven more familiar orders are set out in Table 7·1. Two of the smaller orders are the Trichoptera, or caddis-flies, where both larval and pupal stages are aquatic, building characteristic cases of stones, shells, or plant material and the Siphonaptera, or fleas, where the wings are secondarily absent, the mouth parts adapted for piercing and sucking and their mode of life as ectoparasites on birds and mammals. The order Lepidoptera comprises the familiar moths and butterflies, with two pairs of membranous wings covered with scales, in some cases brilliantly pigmented, and with the adult mouth parts modified to form a long coiled tube for siphoning off free fluids like nectars, while the larvae have biting and chewing mouth parts. Approximately half the known species of insects fall in the order Coleoptera, or beetles, where the fore-wings are modified as elytra which can completely enclose the membranous functional hind-wings when at rest. Several groups of beetles have campodeiform larvae early in development, and several show hypermetamorphosis with additional major changes in structure and way of life occurring within the larval stages.

The order Diptera, or true flies, is another of the largest orders of insects, with the fore-wings functional and membranous, and the hind-wings modified into balancing organs or halteres. The larvae have biting and chewing mouth parts, while the adults have either piercing, sucking mouth parts or a peculiar proboscis adapted for lapping fluids or sponging them up. The blood-sucking forms include mosquitoes, horse-flies, black-flies, and biting midges. Many are important as vectors of tropical diseases, including malaria, sleeping sickness, yellow fever, and elephantiasis. Some species of house-flies, of the genus *Musca* or another, may be the most numerous overt land animals. Finally, another enormous order is the Hymenoptera, comprising ants, bees, wasps, and ichneumon-flies, where the two pairs of membranous wings are interlocked by means of chitinous hooklets. Many are social insects and a large proportion of these show division of labor among several specialized castes. The most elaborate pattern behavior, and most sophisticated methods of communication between individuals of a species, so far discovered in any invertebrates occur in hymenopterous insects.

Representatives of all the other major arthropod groups appear in the geological record earlier than the insects. The earliest fossil insects are from the Carboniferous (Pennsylvanian) Period and include pterygote forms resembling present-day dragonflies and cockroaches. The campodeiform body may reflect an ancient apterygote insect body form, and has some resemblances to the myriapodous minor class Symphyla (see below and Table 2·1).

Minor Arthropods

The four minor classes of arthropods (see Table 2·1) include two which are closely related to the major myriapodous groups, millipedes and centipedes. Their locomotion has already been discussed, but it is worthwhile recapitulating their general features here. The major class, Diplopoda, encompasses the millipedes, predominantly herbivorous animals, moving through deposits of dead leaves and loose soils, and having diplosomites, each bearing two pairs of legs. The major class, Chilopoda, is the centipedes, predominantly carnivorous with relatively long legs and a capacity for fast locomotion. The minor class Pauropoda includes about sixty species of minute, soft-bodied myriapods, probably closely related to the millipedes, and living in forest litter. The minor class, Symphyla, has again about sixty known soil-dwelling species. They superficially resemble centipedes and adults have twelve pairs of legs, insectlike mouth parts, and a pair of spinnerets on the penultimate segment. In some structural features, symphylans resemble the centipedes and in others the apterygote insects.

The remaining two minor classes are probably distantly allied with the arachnids. The class Merostomata comprises two distinct and unequal groups: the extensive fossil subclass Eurypterida, including the

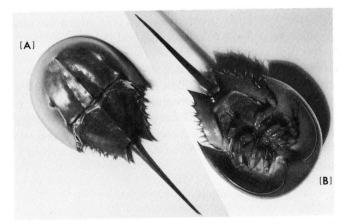

Figure 7·5. A living archetype: the horseshoe crab, *Limulus,* **A** dorsal and **B** ventral views of a young adult. Note the three tagmata, prosoma, mesosoma, and metasoma with telson spine. Of the six pairs of limbs on the prosoma, five are chelate and all have gnathobases, while the limbs of the mesosoma form an operculum and gill-books. [Photos by the author.]

giant water scorpions, and the subclass Xiphosura, encompassing the five species of horseshoe crabs like *Limulus* (Figure 7·5). The horseshoe crabs are marine arachnidlike animals with the body in three tagmata. There is a prosoma with a large semicircular carapace bearing two median and two lateral eyes, and then a hexagonal "abdomen" consisting of the mesosoma bearing six pairs of appendages, including a genital operculum and a series of leaflike lamellae forming the gill-books, fused to the metasoma without paired appendages but bearing the elongate spine of the telson. The paired limbs of the prosoma comprise a pair of small chelicerae and five pairs of walking legs—the first four of which are chelate, bear gnathobases, and closely resemble the single pair of pedipalps in scorpions. Horseshoe crabs are scavengers of the littoral and sublittoral on the Atlantic coast of North America and along the coasts of Southeast Asia. They are the only surviving marine arachnidlike animals, and fossil xiphosurans are known from the Ordovician Period onward, but, unlike other surviving primitive animals, horseshoe crabs are neither rare nor retiring animals. There are few possible predators on *Limulus* except man, and, teleologically, horseshoe crabs seem to regard the rest of the littoral faunas of the regions where they occur with supreme indifference.

The final minor arthropod class is the Pycnogonida, the sea-spiders or nobody-crabs (Figure 8·1C). These are widely distributed marine arthropods, which are never abundant. The trunk is greatly reduced and the four pairs of walking legs usually enormously enlarged. The head bears a proboscis, a pair of chelicerae, and a pair of palps. The abdomen is a tiny vestige, and the alimentary and reproductive systems are largely enclosed in the legs. Detailed study of their segments and appendages suggests some relationships to both arachnids and crustaceans, all much modified by the peculiar body plan. They are dioecious animals, and in many the male broods the eggs until they are hatched. In the majority, the larva which hatches from the egg is termed a protonymphon, which has only three pairs of appendages, though these are quite distinct from the corresponding structures in the nauplius larva of crustaceans. Pycnogonids are browsing carnivores feeding on sessile sponges, coelenterates, and ectoprocts. Without further evidence of their ancestry, the Pycnogonida must necessarily be placed in a separate class of the arthropods.

Arthropod Phylogeny

As already discussed, the great bulk of available evidence suggests that the arthropods evolved from annelidlike animals. Consideration of the relationships between the annelid-arthropod stocks and the rest of the many-celled animal phyla will be postponed to Chapter 14. As we have seen when discussing tracheal respiration and other mechanisms, convergent evolution of structures and functions has occurred to a considerable extent in arthropods. Thus, although there is a unique structural and functional basic plan common to all arthropods, several authorities feel that the group may be of multiple origin from proto-annelid stocks. A diphyletic origin is favored by a number of investigators including Manton and O. W. Teigs. Certain interrelationships of the major classes are more universally accepted. The old association of the Crustacea with the insects and myriapodous groups in the subphylum Mandibulata is almost certainly erroneous and should be discarded, since the mandibles of crustaceans and insects are not homologous in either development or in fossil evidence of their phylogeny. Most authorities agree on a close relationship between the insects and myriapodous groups. There is also agreement that eurypteridlike ancestors probably gave rise to the arachnid groups through intermediate forms resembling the modern scorpions, and that the eurypterids were in turn derived from modified trilobite stocks. The probable close relationship between the Crustacea and the trilobites has already been documented. The proponents of a diphyletic origin of the Arthropoda, including Manton, strongly maintained the hypothesis that the myriapods and insects evolved independently from soft-bodied segmented animals like the Onychophora (see p. 126). However, this would assume that the evolution of a spacious hemocoel, and reduction of the true coelom antedated (as in *Peripatus*) the evolution of the arthropod exoskeleton. As already discussed, there are a number of functional aspects which make this sequence unlikely and also much embryological evidence which suggests that the primitive arthropodan coelom was metamerically divided and probably much more extensive. Those zoologists who prefer the alternative hypothesis that the arthropods form a monophyletic group, and I am among them, must admit that, on presently available data, the insects and myriapodous classes cannot readily be related to the probable stem stock of both arachnids and crustaceans —the trilobites. In spite of this, we feel that the extensive homologies of structure and of function (discussed in Chapters 2, 4, and elsewhere) unite the diverse stocks of arthropods as animals with a uniquely integrated pattern of organs and physiological processes—in other words, as a single phylum.

8

Minor Metameric Phyla

In a relatively natural classification, the largest phylum of animals, the Arthropoda, can be arranged to include ten distinct lines of animals with a chitinous exoskeleton and jointed limbs, as we have seen. There remain four very small groups of animals, metamerically segmented and clearly but distantly related to the annelid-arthropod stock. These are sufficiently distinct from one another and from the two major metameric phyla that each must be considered as a distinct minor phylum. As is the case with the minor pseudocoelomate phyla (see BLI, p. 90), the case for combining such minor groups into superphyla is not a good one. The composite groups formed would be far looser in their relationships than the major phyla and, given our ignorance of their true relationships, it is best to treat them as separate minor groups.

The Walking-worms

From the point of view of broad phylogenies, the phylum Onychophora is certainly the most interesting of the minor metameric phyla. The so-called walking-worms, of which there are about eighty living species and a few fossils, are claimed in many textbooks as the missing links between the annelids and the arthropods (Figure 8·1A, B). They exhibit a mixture of the characters of the two groups, and although in no way really intermediate between modern annelid worms and the Arthropoda, they probably preserve some of the structural and functional patterns of the ancient stocks from which the true arthropods arose. They are found in tropical and south temperate regions and are restricted to highly humid habitats. Like other less well-adapted terrestrial animals, they are cryptic forms, with all the appropriate behavioral reactions.

Although the body surface is covered with a chitinous cuticle, there is no real exoskeleton. Chemically, the cuticle of onychophorans is very permeable and, mechanically, it is thin, flexible, and not divided into a series of articulating plates. Below this are the typical layers of muscles and connective tissues of a wormlike body. However, these muscles surround an enlarged hemocoel, and as Dr. Sidnie M. Manton has elucidated, the body-wall muscles are concerned only in changes of length and of shape of the body, the main processes of locomotion being carried out by the movements of the appendages alone. The speed of locomotion can be modified by changes of gait, and this involves changes in the extension of the body: faster gaits being associated with an elongated body, slower gaits with a shortened one. The paired limbs themselves vary from fourteen to forty-three pairs in different species. Each leg is a stumpy protuberance ending in two claws and with, on the ventral side, from three to six pads which serve as the walking soles which contact the substratum. The body is actually lifted off the ground by the legs which move in a series of steps. In each, the antagonistic contraction of muscles arranged both within the limb and attached to its base, perform in a pattern resembling that of the limbs of primitive myriapods. The head has a series of paired appendages vaguely like those of arthropods but exhibiting no clearly homologous structures. Associated with each pair of walking limbs is a pair of coxal glands, similar to the coelomoducts in annelids, true nephridia being absent. On the other hand, the respiratory organs are tracheae, again independently evolved in this group. Onychophoran tracheae arise from minute spiracles which are scattered all over the surface of

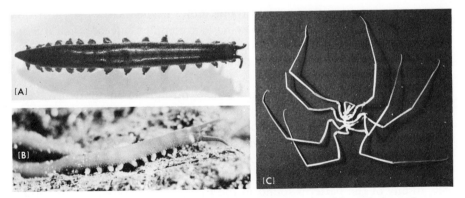

Figure 8·1. Two minor metameric groups. A and **B:** Dorsal and lateral views of a "walking-worm," *Peripatus* (phylum Onychophora). **C:** Dorsal view of a "sea-spider" or "nobody-crab," *Nymphon* (phylum Arthropoda, class Pycnogonida). [**A:** Photo courtesy General Biological Supply House. **B:** Photo by Dr. Joseph L. Simon of a living specimen collected by Dr. Donald J. Zinn. **C:** Photo by the author.]

the body and form a series of tufts of minute tracheal tubules, which do not branch but run as simple tubules to each tissue. The nervous system is essentially annelid in form with the addition of a large bilobed brain, located supraoesophageally, to which are linked the sensory tentacles and eyes. The moderately well-developed sense-organs are obviously important in the cryptic and nocturnal behavior so important to their survival as poorly homeostatic animals on land. The reproductive organs and the relatively large eggs are essentially arthropod in their structure and functioning.

Several zoologists believe that the arthropodan features of the walking-worms are so marked that the group ought to be considered as a subphylum of the phylum Arthropoda. While it seems likely, as Manton and others propose, that forms like *Peripatus* are representatives of the primitive stock from which myriapods and some other arthropods evolved, it seems better to retain a separate phylum for the group. Pragmatically, the diagnosis of the phylum Arthropoda, with its significant functional features, would lose much if stretched to accommodate the onychophorans.

Water-bears or Tardigrades

The phylum Tardigrada consists of minute animals (50 microns to 1 millimeter), with a cuticular exoskeleton divided (in some) into segmental plates and molted during growth in all. There are four body segments, each with a pair of stumpy appendages ending in cuticular claws. There are a few marine species, but most are semi-aquatic, freshwater animals—living in soil water, under or on mosses, lichens, and liverworts. Tardigrades are cosmopolitan and ubiquitous, and it seems to me that they can always be found in the moss which grows on the roofs of old buildings. They have a complex buccal apparatus involving sharp stylets, which are used by all tardigrades to pierce plant cells and suck out their contents. Physiological specialization allows tardigrades to survive in their peculiar habitats.

Although at first sight arthropodlike, tardigrade anatomy involves a jumble of similarities to other phyla, which in most places cannot reflect real relationships. Although there is a molt cycle, the cuticular exoskeleton is not chitinous and does not show the characteristic articulations of arthropods. Development is direct and there are no larval stages. Although the cleavage pattern is holoblastic, it does not conform to any of the patterns found in the major animal phyla. While protostomous in development, tardigrades form five coelomic spaces by enterocoelous means (like echinoderms and chordates!). Only one reproductive coelomic pouch persists, the adult body-cavity being derived from the segmentation cavity and being therefore pseudocoelous. In most cases, subadult growth seems not to involve cell division, and therefore cell number constancy seems to occur (This last feature, along with the pseudocoel, involves similarities to rotifers, Nematomorpha, and the other pseudocoelomate minor phyla [see BLI, pp. 86–90].) Three tubular glands attached to the hindgut have been claimed as excretory organs, but there is little or no evidence for this. There are no circulatory or respiratory organs. The nervous system is obviously metameric, and the muscles show some segmental repetition.

The most important feature of tardigrade physiology is the capacity for diapause. Most of the life of each individual tardigrade seems to be spent in a desiccated, shriveled state. Metabolism continues but is reduced to a very low level in this diapause, which is ended as soon as free water is again available. The animal then swells to four to five times its diapausing volume, and becomes active and feeds on any available plant cells. While in the desiccated state, tardigrades are resistant to fantastic environmental extremes. There are authentic records of tardigrades emerging after dry laboratory storage for seven years and surviving exposure to temperatures of $-272°C$. Individuals in diapause can also be submerged in saturated salt solutions, in absolute alcohol, and even in ether, and survive. The nature of the cuticle obviously requires further investigation, as does the molt cycle, which, however, is clearly different in many respects from that in arthropods.

Tardigrades have separate sexes, internal fertilization, and moderately large eggs. It seems that tardigrades, like rotifers and gastrotrichs, can produce eggs of two sorts: those for immediate development, thin-shelled, and those resistant to desiccation and diapausing. It is possible that eggs are also produced parthenogenetically as in rotifers and cladocerans. Although most textbooks give figures of from three hundred to four hundred described species for the phylum Tardigrada, and postulate that many more remain to be discovered, this seems to result from excessive "splitting" and there are probably only about one hundred and seventy extant species. It is characteristic of cosmopolitan freshwater groups relying on passive dispersal that there is a very high degree of infraspecific interpopulation variation and, conversely, that species and genera are relatively widespread and ubiquitous.

One last feature of tardigrades deserves to be mentioned. They have an alluring charm for certain biologists, out of all proportion to their importance. Many invertebrate physiologists have revealed to me their dreams of a profitable series of researches on tardigrade functions—a vision I share. Perhaps an hour spent at a microscope watching some tardigrades come out of diapause would enroll any student of biology in this company of dreamers.

Phylum Linguatulida

The pentastomids, or linguatulids, form a minor phylum of about sixty species of wormlike, parasitic animals. Once again, there seems to be some relation to a pre-arthropod stock. The body is covered by a thick chitinous cuticle and there is a regular molt cycle during larval development. They are found in the lungs and respiratory passages of certain reptiles, birds, and mammals. The host of the adult is always carnivorous, and in most linguatulids the life-cycle requires an intermediate host for the larval development. In some forms the larva resembles a tardigrade. There are no respiratory, circulatory, or excretory organs, but the nervous system is metameric, with the annelid-arthropod form. Like the onychophorans, the linguatulids seem to derive from an unknown pre-arthropod stock.

Phylum Echiuroidea

The echiurids form a minor phylum of marine animals which have obvious resemblances to annelid worms, but which show no trace of metamerism as adults, and which are characterized by the possession of an enormous nonretractable grooved proboscis which is used in feeding. There are more than eighty species, and the majority live in shallow water of the sublittoral.

The body wall is annelidlike and there are setae which are chitinous and formed in sacs as in earthworms. Development involves spiral cleavage and a typical trochosphere larva leading to a metamerically segmented embryo. Almost all traces of metamerism are then lost during later development. The excretory system and circulatory system are like those in primitive annelids, although the central nervous system is much simpler.

On the other hand, some features are distinctly unlike the annelid worms. The coelom is an open cavity without septa and the gut is a very long, convoluted tube. The diagnostic feature of the group is the large, extensible grooved proboscis which is actually a cephalic lobe corresponding to the prostomium of annelid worms, and containing the brain. The edges of the proboscis are rolled to form a gutter which is ciliated and this is used in obtaining food. Most genera are detritus-feeders, catching food particles on mucus secreted by the proboscis and carrying it back to the mouth along the ciliated groove. The Californian genus *Urechis* constructs U-shaped burrows and spins mucous nets to catch particulate food. These mucous nets can apparently be spun with a pore size fine enough to stop protein molecules of molecular weight about five hundred thousand, while passing those of molecular weight around one hundred thousand. This filtration of protein molecules and other small particles does not seem to involve adsorption onto the mucus. Several of the detritus-feeding species can extend the proboscis to ten times the length of the trunk, that is, for distances of about 1 meter.

This minor phylum seems to have arisen from a group ancestral to the present-day Annelida, in which metameric segmentation had already been evolved.

9

Larval Types, and Further Minor Phyla

IT IS NOT SURPRISING that writers of zoological textbooks have often tried to set up assemblages of "related" phyla, in an attempt to reduce the thirty or so animal phyla (Table 1·1, p. 5) to two or three "major evolutionary stocks." Further, it is not surprising that characteristics of embryonic and larval development have been used in the logical diagnoses of "superphyla," "animal subkingdoms," or other higher taxa of doubtful significance.

Biology as a science has always depended on the perception of fundamental unities within the prodigious variations of nature. As regards the study of invertebrate animals, this statement is so obvious as to be trite. Classical studies on comparative anatomy revealed the unity of basic structural pattern in each phylum. Although varying in external characteristics, the members of a typical phylum form an assemblage, all constructed on the same ground plan with certain essential groupings of structural units. A major part of the *raison d'être* of this book and of its companion volume (BLI) is that the discussion of similar homologies of function and the use of working archetypes—each involving an integrated functional plan, in which the concert of organs and functions can operate as a whole—are of value in attempting to comprehend the diversity of form and function in invertebrate animals. However, as stressed elsewhere (see pp. 4 and 206, and BLI), *real* ancestry is ultimately unknowable, and statements on homologies, archetypes, and phylogenetic classification—including all comments on the interrelationships of animal phyla—are matters involving hypotheses. The difference between the "hard" data on patterns of development and the subjective inferences therefrom regarding questions of degrees of interrelationship and common descent should be remembered during the discussion which follows. The phylogenetic value of embryological data was established in the nineteenth century, but modern biology has taught us that the interpretation of developmental evidence is rarely as straightforward as von Baer and Haeckel proposed. It is most important to realize that the *special requirements of embryonic and larval life* are responsible for adaptive modifications involving the structures and time-pattern of development. Features exhibited by larval types can reflect *either* recapitulation of ancestral history *or* adaptational response to the immediate needs of larval life.

Patterns in Embryos and Larvae

Many, perhaps the majority of, textbooks of zoology divide the more complex phyla of many-celled animals into two great assemblages: the Protostomia and the Deuterostomia. Platyhelminthes, Mollusca, Annelida, Arthropoda, and a number of minor phyla are classified as protostomes, while the Echinodermata, the Chordata, and at least two minor phyla are included in the deuterostomes. (A few texts even include the Cnidaria and Ctenophora among the protostomes, but this is a ridiculous extension based partly on misconceptions).

The features claimed as diagnostic of this division are largely developmental. The Protostomia are said to show: spiral cleavage of a mosaic egg, the mouth formed from the blastopore, the body-cavity schizocoelous in origin, the central nervous system ventral and deep-seated, and, if a free-living larva is formed, a trochospherelike larval organization. The Deuterostomia are said to show: radial and indeterminate cleavage, the mouth formed as a secondary opening with the blastopore becoming the anus, the body-cavity enterocoelous in origin and usually showing tripartite subdivision, the central nervous system superficial and usually dorsal, and, if a free-living larva is formed, a "dipleurula-like" organization. Although based on facts of development, unfortunately the five diagnostic features are not nearly so consistently associated together as some authors suggest. Further, it is impossible to make a clear dichotomy of animal phyla on the basis of these five characteristics. Some of the protostome-deuterostome distinctions, where they occur, may reflect phyletic relationships, but they are neither so universal nor so consistent as to justify being used in major classification. However, it is worth examining in a little more detail some of the factual data of development involved. Perhaps the most significant developmental data concerns the occurrence of spiral cleavage. The eggs of annelids, molluscs, nemertines, and the polyclad platyhelminths show several patterns of this form of cleavage. Spiral cleavage is mosaic in the classical sense, that is to say that each blastomere formed has a predetermined and fixed fate in the later embryo. With this determinate cleavage, experimental embryologists cannot produce twins from one fertilized egg. The contrasting indeterminate radial cleavage is seen best in some echinoderm eggs. In these, if the cells of the four-cell stage are artificially separated, each is capable of forming a complete gastrula and then a young larva.

Among the arthropods only some barnacles show a form of spiral cleavage and a variety of patterns of early development are found in the minor phyla. As regards the embryonic origins of the mouth, in some protostomes the blastopore becomes the mouth while in others the blastopore closes and a new mouth opens later close by. The origins of the anus in deuterostomes vary similarly. In many protostomes, the coelom arises as a schizocoel, by splitting of the mesoderm layer. The coelom in deuterostomes forms by a process called enterocoelic pouching in which the wall of the archenteron itself evaginates to form the mesoderm. The processes of coelom-formation are not really consistent with the other characters. For example, the Brachiopoda have radial cleavage, protostomous mouth formation, and an enterocoelous body-cavity. Indeed, taken as a group, the lophophore-bearing phyla (Ectoprocta, Phoronida, and Brachiopoda) are in many respects intermediate between "true protostomes" and "true deuterostomes." There are other aberrations of pattern: the tripartite division of the coelom supposedly characteristic of the deuterostomes is also found in the Chaetognatha, Tardigrada, and other unrelated minor phyla.

Finally, we have the evidence provided by the free-swimming ciliated larvae where they occur. A trochosphere larva is shaped like a spinning top with a tuft of cilia on the apex and an equatorial band of cilia through which the mouth opens (see Figure 2·2A; see also BLI, pp. 102–103). Typical trochosphere larvae occur during the life-cycle of primitive polychaetes and archiannelids. These larvae seem well-adapted for feeding and locomotion in a planktonic environment and retain their characteristic form until the onset of metameric segmentation (see pp. 12 and 21). A closely similar larva, the trochophore (see BLI, pp. 123, 165), is found in many molluscs, where it develops into the more characteristic veliger (see BLI, pp. 123 and 125 and Figure 9·6). The flatworms and nemertines have markedly different ciliated larvae, as do most of the minor protostome phyla, although the Entoprocta have a "trochosphere." The characteristic larva of the deuterostomes has been claimed to have a "dipleurula" form, but this involves a synthesis from two distinct types of actual larval organization: the pluteus and the bipinnaria-auricularia (see Figures 11·4 and 11·5). Strictly, these larvae, which are clearly distinct from the trochosphere form, are characteristic of the phylum Echinodermata alone. No true chordate has such a larva, and the peculiar case of the hemichordate tornaria will be discussed later (see p. 189). Like the trochospheres, plutei and other free echinoderm larvae are well-adapted for continuous locomotion and feeding on microflagellates in the plankton.

Obviously, many of the difficulties encountered in attempting to erect superphylum assemblages arise from the variety of functional patterns found in the twelve minor coelomate phyla. Of these, the four with obvious annelid-arthropod relationships were discussed in the last chapter, and two with some connection with the chordates are discussed in Chapter 12 below. There remain six phyla, including the lophophore-bearers, and these will be considered now.

Phylum Ectoprocta

The ectoprocts are the coelomate moss-animals, with a superficial resemblance to the phylum Entoprocta, which are totally distinct and pseudocoelomate (see BLI, p. 86). The microscopic, individual ectoprocts, termed zooids, live in extensive, sessile colonies, each fastened within a secreted exoskeletal box, or zooecium, and feeding by means of an extensile lophophore bearing ciliated tentacles (see Figures 9·1 and 9·2). Many genera, like *Elektra,* form creeping mats while others produce upright branching systems reminiscent of hydroid coelenterates.

Although individuals are microscopic and colonies never massive, it may be incorrect to speak of the Ectoprocta as a minor phylum. There are probably about four thousand living species and possibly about fifteen thousand different extinct forms. On the other hand, they are hardly a major group in the ecological sense discussed earlier in this

Figure 9·1. Colony of a freshwater ectoproct, *Lophopodella carteri,* showing the lophophores in their expanded feeding position. [Photo by Dr. Shuzitu Oda (1960), courtesy of Dr. Thomas J. M. Schopf.]

book, and only make up a minute fraction of the biomass in the marine and freshwater habitats where they occur. Appreciation of the ectoprocts is hampered by there being little in the way of good biological studies on living zooids, but an enormous literature on the details of the architecture of their boxlike houses, the zooecia.

Feeding in ectoprocts is carried out by the lophophore, which may be horseshoe-shaped or circular. When fully expanded, the lophophore forms a funnel of diverging tentacles surrounding the mouth and leading into it. Each tentacle is ciliated with two lateral tracts of cilia (facing those of other tentacles) and a third tract of shorter cilia on the median inner surface. It is not clear whether trapped food particles are carried continuously down these tentacles to the mouth while the lophophore remains expanded, or whether retraction of the lophophore is necessary to carry the captured food into the mouth. In all ectoprocts the lophophore can be rapidly contracted into a bundle of tentacles and pulled down into the zooecium. The gut is U-shaped but the anus lies outside the lophophore. The trunk of the body has a peritoneum-lined coelom separating the gut from the body wall, and this coelomic cavity extends into the lophophore and tentacles which are extended by various hydraulic means. Extension and retraction of lophophores in ectoprocts having different forms of zooecia involve different elaborations of the musculature. There is always an important group of lophophore-retractors usually inserted on the coelomic septum at the lophophore base and originating on the aboral wall of the zooecium. Protrusion of the lophophore is carried out in a variety of ways, all esssentially hydraulic. Some simpler forms of zooecia have the frontal wall (that is the roof of the box) made of a flexible membrane to which muscles are attached. Contraction of these muscles decreases the volume of the box, thus compressing the coelomic fluids of the zooid and extruding the lophophore. In some more complex forms, the frontal membrane still provides the compression to extrude the lophophore, but a perforated calcareous shelf, termed the cryptocyst, forms a second roof below the flexible one and serves for the better protection of the zooid. In yet another group of ectoprocts, the Ascophora, the zooecial box is completely calcified. In these forms a thin-walled sac, the compensation sac or ascus, has been formed inside the zooecium and has a separate opening through a pore to the exterior. In this case the contraction of muscles attached to the floor of the ascus brings about its dilation, thus sucking seawater in from the exterior. It also causes compression of the coelomic fluid in the zooid and thus extrusion of the lophophore (Figure 9·2).

Clearly the lateral cilia of the tentacles move water downward and centrifugally but there do not seem to be ciliary tracts leading directly to the mouth except round the lophophore base. Large unwanted particles can be rejected by muscular flicking of the tentacles or by total

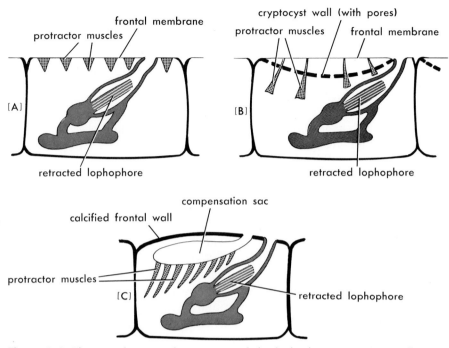

Figure 9·2. Three mechanisms of protraction of the lophophore in ectoprocts. For further explanation, see text.

retraction of the lophophore. There are cilia in the pharynx and throughout most of the gut, and in some ectoprocts the food is formed into a mucus-cord which is rotated as it passes through. It seems that in the commonest marine ectoprocts (class Gymnolaemata) digestion is partially extracellular and partly intracellular, as in the coelenterates and many molluscs. As in many other filter-feeding organisms, the hindgut is concerned in consolidation of the faeces which in ectoprocts are formed into mucus-covered balls and expelled away from the feeding mechanism of the lophophore.

The nervous system forms a plexus in the body wall with discrete nerves running into individual tentacles and around the lophophore, and a main ganglion between the mouth and the anus (which position is termed dorsal). Not unexpectedly, in view of the minute size of ectoproct zooids, there are no circulatory, respiratory, or excretory organs. Ectoprocts are mostly hermaphrodite, with the gonads developed on the peritoneum. In about 90 per cent of the ectoprocts studied, fertilized eggs are retained in a brood-pouch. In the few forms where the ova develop externally a highly modified form of trochosphere larva— the cyphonautes larva is formed. This, or an even more modified larva, released from the brood-pouch, settles on a hard substratum, undergoes a degenerative metamorphosis, and forms the founder zooid of a new colony. Colonies are of course produced by asexual budding and in different ectoproct species there are different degrees of continuity between individual zooids in the colony. A few are polymorphic. In the freshwater forms there is also asexual production of balls of cells termed statoblasts which are like the gemmules of sponges and which can survive desiccation and freezing to develop in suitable spring conditions. Once again, as in rotifers (see BLI, pp. 87–88), the success of a relatively minor group of minute animals in temperate fresh waters depends on a modification of the life-cycle to suit the peculiar conditions of that environment.

As mentioned earlier, much of the detailed classification of ectoprocts depends on the characters of the zooecia which may be calcareous, gelatinous, or membranous—some of the latter being formed of chitin. The phylum is naturally divided into two classes. The Phylactolaemata are the ectoprocts of fresh water with the lophophore usually horseshoe-shaped. In these, the coeloms of the zooids are continuous through the colony, the zooids never show polymorphy, and the zooecia are never calcareous. Remarkably, and by unknown mechanical means, colonies of the freshwater genus *Cristatella* are motile. The other and much larger class, the Gymnolaemata, includes all marine ectoprocts plus a few in fresh and brackish waters. In this class the lophophores are circular, there are no coelomic connections between zooids, there can be polymorphy, and there is a wide variety of zooecial architecture.

Phoronids and Priapulids

The minor phylum Phoronida—there are only about fifteen species—consists of wormlike animals with a lophophore. Although they share the possession of a lophophore with the Ectoprocta and Brachiopoda, the phoronids are totally unlike these other groups in external anatomy and way of life. The small number of known species, which can probably all be assigned to the genera *Phoronis* and *Phoronopsis,* all live in the marine littoral as solitary individuals in secreted tubes, though many of these tubes may be aggregated together. Each tube is a permanent residence, formed of secreted chitinous material coated with pebbles and shells. Only the lophophore is thrust out for feeding and it can be rapidly retracted. The gut is U-shaped with an anus near the head, and there is a pair of nephridia with pores on either side of the anus. The coelom is subdivided with a separate lophophore cavity and there is a circulatory system with hemoglobin in corpuscles circulating in two longitudinal vessels with a ring vessel connecting them anteriorly. The nervous system consists of a diffuse net with a ring around the base of the lophophore and one giant nerve fiber running down the left side of the trunk, which functions to bring about the quick syn-chronous contraction which brings in the lophophore.

Some phoronids are hermaphrodite and some dioecious, but most brood their young for some time at the base of the lophophore. The larvae eventually released are the characteristic and peculiar actino-trochs, which some zoologists claim resemble the trochophore. Other phylogenies suggest that the phoronid worms are related to the stock which gave rise to the deuterostome phyla.

The priapulids are another minor group of marine wormlike animals. Only five species have been described. Their treatment in this chapter is somewhat arbitrary since, among other difficulties, it is not clear whether they are coelomate or pseudocoelomate animals. In many anatomical features, they are intermediate between the coelomate phylum Sipunculoidea and the pseudocoelomate phyla, including the rotifers and acanthocephalans. Like the sipunculids, the body is di-vided into a proboscislike presoma and a trunk. This presoma is re-tractable and bears the mouth which is surrounded by a toothed cuticle whose spines are arranged in a series of concentric double pentagons. (Although no relationship is likely, echinoderms show similar cir-cumoral pentaradiate symmetry.) Priapulids are thought to be car-nivores capturing slowly moving prey with the circumoral spines and swallowing it whole. The gut is a straight tube running to a terminal anus, that is, quite unlike its condition in the sipunculids. There are protonephridia made up of large numbers of nucleated solenocytes. The gonads form tangled masses of tubules. Early cleavage stages are radial, not spiral. The larva which results has the cuticle in the form of plates like the lorica of a rotifer, and the adult form is acquired through a series of cuticular molts.

Perhaps the most peculiar feature, and one whose significance is not yet clear, is that there is a spacious open body-cavity. This is lined with a thin noncellular layer which also covers the viscera and forms double sheet slings like the mesenteries of true coelomates. If it were not for its noncellular nature, this would be regarded as being homologous with the peritoneum in groups like the Annelida. In other ways the priapu-lids resemble pseudocoelomates and early this century would have been incorporated in the now-discarded superphylum Gephyrea (see BLI, p. 90).

Phylum Brachiopoda

The lampshells, phylum Brachiopoda, again illustrate the phyletic unreality of assemblages involving combinations of animal phyla. In zoological textbooks where such groupings are used, the brachiopods

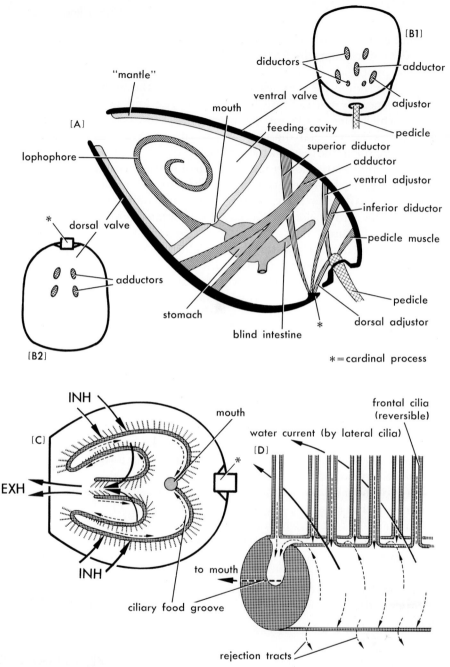

Figure 9·3. The functional organization of a brachiopod. A: Vertical section through a typical testicardine brachiopod, showing the elaborate systems of adjustor, adductor, and diductor muscles and the position of the lophophore within the feeding-cavity. **B1** and **B2:** Ventral and dorsal valves, showing the muscle insertions. **C:** Major water currents employed in feeding, in relation to the coiled arms of the lophophore. **D:** Detail of water currents and ciliation of surfaces on part of a lophophore arm. For further explanation, see text.

are always placed among the protostomous phyla *but* their coelom is formed in the enterocoelous manner.

Lampshells show a superficial resemblance to bivalve molluscs, though the two valves of the brachiopod shell are always attached dorsal and ventral to the soft parts (shell valves in Bivalvia are lateral), and the feeding organ is a complex circumoral lophophore. They are all marine animals and the three hundred or so species presently living are merely a remant of a once dominant faunal group. The living brachiopods are placed in sixty-eight genera while the fossil genera described from Palaeozoic and Mesozoic seas number in excess of one thousand seven hundred. Various authorities place the number of fossil species between twelve thousand and thirty thousand—they were undoubtedly abundant animals and are found as the dominant fossils in many limestone deposits. An extremely complex classification has been set up, but only two main divisions need be mentioned in relation to the presently living forms. The Ecardines are brachiopods whose valves are united by muscles only, without an elaborate hinge, and whose shell is largely calcium phosphate. They lack an internal skeleton for the lophophore and have a gut with an anus. As adults, they may be cemented down to the substratum like oysters (*Crania*), or they may have the shell valves borne on a very long pedicle as in *Lingula*. The latter is an extreme example of a persistent genus, being found in deposits from the Cambrian to the present—apparently with no change in the anatomical features preserved.

In contrast, the Testicardines (or Articulata) have hinged structures uniting the shell valves, an internal skeleton supporting the lophophore, always a short pedicle through an opening in the ventral valve, a gut without an anus, and a shell largely composed of calcium carbonate (Figure 9·3A). Typical genera such as *Terebratula* and *Waldheimia* live mostly in the deeper parts of the oceans.

As with the clams and oysters of the phylum Mollusca, there is a certain uniformity of anatomy and physiology which is imposed in sessile animals using a ciliary feeding mechanism. In lampshells, the feeding mechanism depends on water currents created by the cilia of the lophophore. (It is important to realize that although there are many parallels to the bivalve molluscs, the use of terms like mantle, lateral cilia, and so on does *not* imply any homologies with the molluscan systems—see BLI, pp. 135–140.) Lobes of the body wall, termed the mantle, secrete the shells and enclose the feeding cavity with its lophophore. The mouth at the center of the lophophore opens, by way of an oesophagus, to a stomach with one or more pairs of digestive diverticula and a blind intestine (or one which opens to an anus in the part of the feeding-cavity with an exhalant water current). There is an elaborate system of muscles which move the shells relative to each other and to the pedicle and also adjust the lophophore, their pattern varying from group to group. Brachiopods are coelomate, with an open circulatory system consisting of a mid-dorsal vessel above the gut with open branches at each end. They may or may not have contractile vesicles in this system. There are one or two pairs of metanephridia which also act as gonoducts and may well be nephromixia. Brachiopods are mostly dioecious and spawn or retain eggs in the feeding cavity to shorten the free-swimming larval life. The larva is trochosphere-like !

When the animal is feeding (see Figure 9·3C), the valves gape and there is an inhalant current from the sides and a median exhalant current. The main cilia creating this current are on the sides of the tentacles along the lophophore and are termed the lateral cilia. It is important to note that the double row of tentacles are placed so as to alternate on either side of the brachial groove (see Figure 9·3D). Food is strained off by the ciliated tentacles and carried by the cilia on their faces into the brachial groove and along this groove on its circuitous route (see Figure 9·3C) to the mouth. These cilia on the faces, or frontal cilia, however are reversible and can be adjusted to reject unsuitable food particles to the tips of the tentacles and off these onto the cavity wall and out in the exhalant current, thus providing an almost exact analog of the rejection of unwanted particles by bivalve molluscs as pseudo-faeces.

There is another origin for rejected material. This lies in the tracts of cilia on the underside of the lophophore, that is, on the opposite face to the brachial groove. As shown in Figure 9·3D, particles impinging on the lophophore in this area of its surface are not carried to the mouth but again rejected onto the walls of the feeding-cavity. It is perhaps one of the points where the efficiency of brachiopods is markedly less than that of bivalve molluscs, that lampshells do not seem to have evolved any elaboration of ciliary tracts for the later disposal of pseudo-faeces, nor any neuromuscular patterns resembling the clapping of bivalves, to carry out the further disposal of pseudofaeces. Apart from this, however, lampshells are seemingly relatively efficient machines for ciliary filter-feeding, and it is very tempting to suggest that their decline in importance in the fossil record parallels the increasing abundance of even more efficient filter-feeding machines—the bivalve molluscs such as mussels, clams, and oysters.

The details of structure and the finer aspects of functioning in lophophores are both more easily studied and already better described in brachiopods. More comparative studies could establish whether or not the lophophores in the other kinds of animals which bear them, including the Ectoprocta and Phoronida, are structurally and functionally homologous with those in brachiopods. Obviously, this would be of considerable phyletic significance.

The minor phylum Sipunculoidea is rather distinct from the previous groups, and indeed from most other phyla. Some zoologists have tried to link these unsegmented, wormlike animals to the Priapulida, and thence to such acoelomate minor groups as the Rotifera and Gastrotricha, but this is indefensible. Though externally similar to the Echiuroidea, which are probably related to annelid worms (see above, pp. 128–129), there are no traces of metameric segmentation nor of setae. The sipunculids are unsegmented worms with the anterior third of the body as an introvert and with a fundamentally U-shaped gut running to an anus near the anterior. This alone would suggest some relationship to the other more sessile minor coelomate groups such as the ectoprocts and phoronids. There are nearly three hundred species of sipunculid worms, all marine, but worldwide in distribution. A typical genus, *Golfingia,* is found burrowing in intertidal, muddy-sand flats, in temperate seas. (These include the muddy shores off the "Royal and Ancient" golf-course at St. Andrews, Scotland, and the generic name results from an encounter of W. C. McIntosh and E. Ray Lankester.) As in other worms, the muscles of the body wall in *Golfingia* are organized around a hydrostatic skeleton of coelomic fluid. Extremely high coelomic pressures can be generated when grossly overstimulated, and moderately high coelomic pressures (of the order of 200 centimeters of water) are normally developed during protrusion of the introvert. In normal burrowing locomotion, however, the coelomic pressures generated are an order of magnitude less. The burrowing cycle consists of everting the introvert into the substrate, expanding its end to form a mushroom anchor, and then pulling the body up to this by contraction of longitudinal body-wall muscles. It is likely that the highest coelomic pressures are normally only associated with faster escape reactions.

It is of some ecological and evolutionary interest that this representative of a minor phylum, *Golfingia,* commonly lives in the substrate of mudflats with a wide variety of other wormlike animals, and in probable competition with some of them. These include, besides a variety of annelid worms, burrowing sea-anemones, representatives of the Rhynchocoela and even of the Echinodermata (the wormlike holothurian, *Leptosynapta*).

The arrow-worms of the marine plankton form the minor coelomate phylum Chaetognatha. Although clearly deuterostomous in their early development and having a modified enterocoelous development of the body-cavity, chaetognaths do not resemble the other deuterostomous phyla in any way in their adult anatomy. Adult arrow-worms even lack a peritoneum lining the coelom. The phylum is clearly a minor one with sixty or so species, but local abundance of arrow-worms can be of considerable importance in the ecology of the marine plankton (Figure 5·7). The majority of the species in the phylum are placed in one genus, *Sagitta*. Coelomate and unsegmented, they have transparent, torpedo-shaped bodies showing perfect bilateral symmetry. The mouth is sur-rounded by a group of grasping spines, and arrow-worms are predaceous on other members of the zooplankton. They have horizontal fins on sides and tail, and pursue their prey with surprisingly rapid darting movements. The body-wall musculature responsible for their swift swimming consists of contractile elements with undifferentiated cell bodies somewhat similar to those found in nematode worms. In-deed, this undifferentiated musculature and the lack of an appropriate coelomic lining make the body organization of chaetognaths surpris-ingly similar to that of the successful pseudocoelomates. However, chaetognath development makes it clear that this cannot be other than convergence of structural organization. The nervous system consists of a pair of cerebral ganglia associated with a pair of eyes, and a large ventral trunk ganglion. The digestive system consists of a straight tube from the mouth to the anus near the tail and it is surrounded by a coelom with septal divisions separating it into head, trunk, and tail compartments. The trunk section is usually laterally paired and the tail section may be. Some comparative zoologists have claimed that this shows more than accidental resemblance to the three- or fivefold system of divisions of the developing coelom shown in the phylum Chordata and its assumed allies. There are apparently no organs or tissues specialized for excretion, circulation, or respiration. Arrow-worms are hermaphrodite with a pair of ovaries and a pair of testes. Self-fertiliza-tion is possible. The eggs are planktonic and, as already mentioned, development involves radial cleavage, gastrulation by invagination, and is deuterostomous and enterocoelous.

Although there are few species of arrow-worms, they can be among the commonest planktonic animals. They are often present in enormous numbers with significant effects on the marine economy, for example, by the destruction of whole broods of eggs or young stages of some of the commercially important fishes. Specific *Sagitta* are important as planktonic indicators of oceanic waters of different origins. Bodies of water which differ only very minutely in their physical and chemical characteristics are often typified by having, as biological indicators, specific arrow-worms. In many cases, these biological indicator species are more useful than microanalyses to oceanographers interested in the proportionate origins of bodies of oceanic water. A classical example of this is found in the temporal changes in the hydrography of the North Sea. To summarize a fairly complicated relationship, the presence of

Sagitta elegans replacing *Sagitta setosa* is an indication of considerable influx of water from the northern Atlantic, and usually occurs in late summer and early fall of each year. Changes in micronutrients which accompany this influx of oceanic water are of great importance to the productivity of the area. Thus a series of census of *Sagitta* opens up a real possibility of detailed prediction of the returns from herring fisheries in the area.

CHAPTER 10

Echinodermata I: Discerning a Unique Archetype

THE ECHINODERMS form the most easily recognized phylum of animals. This has two implications. They show no clear relationship with any other phylum, except for the distant association with the chordates which will be discussed later. A second implication is that the features considered diagnostic of the phylum are universally possessed, and are both obvious and of functional importance.

The diagnostic feature most often stressed in textbooks for this group is the possession of radial symmetry with a pentaradiate basis. Two other unique features of the peculiar anatomy of echinoderms would seem to be of even greater physiological significance. Indeed the functional potentialities (and limitations) of the phylum Echinodermata would seem to hinge on the possession *both* of a water-vascular system (as a subdivision of the coelom) and of the skin ossicles which give the phylum its name. To anticipate later details, animals like sea-urchins and starfish are moved by numerous tube-feet (hydraulically operated by the water-vascular system), each of which is mechanically supported by passing between the numerous ossicles which closely approximate, or through a pore in one of them. It is difficult to visualize any way in which these many tiny locomotor units could create a gross movement of the starfish relative to the substratum, unless their muscles were inserted on and through such a skeletal system of dermal ossicles. Another of the more unusual characteristics of echinoderms results from the fact that the interlocking system of ossicles (in forms like sea-urchins) consists of units which have to be enlarged during growth. This results in patterns rather more rectilinear than are found in most other living organisms. A glance at the test of a sea-urchin (or at Figure 11·3B) will show the nature of this rectilinearly integrated substructure.

The starfish, sea-urchins, brittle-stars, and the like which make up the phylum (see Table 10·1) are entirely marine, but they are widely distributed throughout the seas and are a successful group by any measure. There are also huge numbers of echinoderm fossils and several groups of them have been established as successful marine animals over long periods of geological time.

They are triploblastic, coelomate animals (see BLI, and pp. 6–7), and show no trace of metameric segmentation. The five- or tenfold radial symmetry which they show as adults is diagnostic, although most echinoderms are bilaterally symmetrical during their larval development, and it would seem that the group shows secondary radial symmetry derived from originally bilaterally symmetrical ancestors. This will be discussed more extensively in the next chapter. Echinoderms have no heads, nor do they show an anterioposterior axis. The nervous system is without a brain and remains superficial throughout life, much of it in contact with the ectodermal epithelia from which it is derived. There is a calcareous endoskeleton of ossicles in the dermal layers (in some echinoderms this makes up the bulk of the animal), and these often are associated with external spines or bosses. The coelom in the adult supposedly derives from three divisions (or more correctly three pairs of divisions) in the embryo. In the adult echinoderm it is anatomically much subdivided, and of varied functional significance. One part is the water-vascular system with its associated external podia (which can function as tube-feet in locomotion, or as tentacles in a food-collecting mechanism). Other subdivisions of the coelom surround the general viscera, and are associated with organ systems such as the so-called "hemal" system. The strands of lacunar tissue which make up this latter system are each enclosed in a tube of coelomic origin. Its anatomical layout (see Figure 10·2A) resembles that of the water-vascular system and of several other organ systems. In all cases there is a circumoral element with five radial elements diverging from it, and thus the diagram of the layout of any system is similar to all the others. There being no head or bilateral axis of symmetry, the terms dorsal and ventral, anterior and posterior, are completely inappropriate. Anatomically there are two surfaces, which do have considerable physiological significance. One surface is the oral or ambulacral—bearing not only the mouth but the ambulacra or areas where tube-feet protrude from the surface. The other surface is termed aboral, and never has external projections of the water-vascular system as podia of any type. Of the five common living classes of echinoderms, four are more motile and keep their oral surfaces down to the substrate (see Figure 10·1). The fifth group, the crinoids (and perhaps many extinct groups of echinoderms), have the aboral surface directed downward and the mouth facing up. As shown in the outline classification (Table 10·1), which incidentally overstresses the comparative importance of the living groups, the orientation of the surfaces corresponds to the major subdivision of the phylum into the Eleutherozoa (including the four motile classes), and the Pelmatozoa (including several fossil groups and the Crinoidea). It should be noted that this outline, and our treat-

TABLE 10·1

Outline Classification of Phylum Echinodermata

Subphylum III ELEUTHEROZOA
   Class 6 SOMASTEROIDEA (only one genus living, *Platasterias*)
   Class 7 ASTEROIDEA (at least 5 orders living) "Starfish," "Sea-stars"
   Class 8 OPHIUROIDEA (at least 2 orders living) "Brittle-stars"
   Class 9 ECHINOIDEA   "Sea-urchins"
     "REGULARIA" (at least 2 orders living) "Sea-urchins," "Sea-hedgehogs"
     "IRREGULARIA" (at least 2 orders living) "Heart-urchins," "Sand-dollars"
   Class 10 HOLOTHUROIDEA (7 orders living) "Sea-cucumbers," "Trepangs"
   Class 11 HELICOPLACOIDEA (fossil only)
   Class 12 OPHIOCISTIOIDEA (fossil only)
Subphylum I HOMALOZOA
   Class 1 CARPOIDEA (fossil only)
Subphylum II PELMATOZOA
   Class 2 CYSTIDEA (fossil only)
   Class 3 BLASTOIDEA (fossil only)
   Class 4 CRINOIDEA (at least 2 orders living) "Sea-lilies," "Feather-stars"
   Class 5 EDRIOASTEROIDEA (fossil only)

ment of the living forms here, inverts most phylogeny, since the Pelmatozoa and Homalozoa are undoubtedly more ancient groups.

It should be realized that considerable controversy surrounds the classification of the echinoderms. The treatment adopted here, and in Table 10·1, represents a compromise, and may thus prove unacceptable to all authorities. Into the most usual arrangement of the living classes (that adopted by Th. Mortensen, Libbie H. Hyman, and many others) is inserted the class Somasteroidea, which in some ways forms a "link" group between the Eleutherozoa and the Pelmatozoa. This results from the work of H. Barraclough Fell, who has clearly elucidated that this group, with one living representative, *Platasterias,* exhibit characters intermediate between those of crinoids and the other starfish. On the other hand, the complete rearrangement of classes proposed by Fell, and involving rejection of the two living subphyla, Eleutherozoa and Pelmatozoa, has not been adopted here. Fell has proposed a division into four subphyla based largely on the organization of growth gradients in adult forms (i.e., distinguishing between those with meridional gradients and those with radially divergent gradients). Some of the phyletic aspects of this will be discussed in the next chapter (see pp. 171–173), but it is worth mentioning here that authors utilizing the customary classification (including Hyman) have long admitted that the Eleutherozoa probably do not represent a monophyletic grouping. In other words, it is freely admitted that the more motile echino-

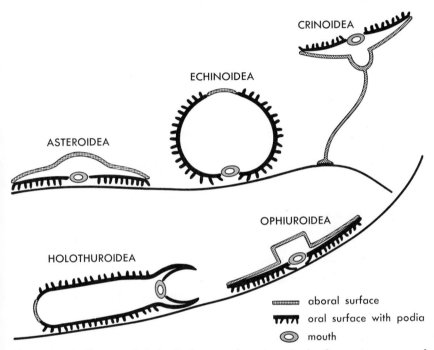

CRINOIDEA

ECHINOIDEA

ASTEROIDEA

OPHIUROIDEA

HOLOTHUROIDEA

▭▭▭▭ aboral surface

ⱱⱱⱱ oral surface with podia

◎ mouth

Figure 10·1. Characteristic body form and posture in the five major groups of living echinoderms: Asteroidea (starfish or sea-stars), Echinoidea (sea-urchins), Crinoidea (sea-lilies and feather-stars), Ophiuroidea (brittle-stars), and Holothuroidea (sea-cucumbers or trepangs).

derms have originated from the more ancient crinoidlike ancestors along at least two differing lines. Inclusion of a separate subphylum Homalozoa to encompass the "carpoids" again departs from the "usual" classification. This is certainly justified (even in a neontological account) since the fossils of the class Carpoidea could represent the most ancient echinoderms. Found in Palaeozoic deposits, they are bilateral with dorsoventral flattening and show no evidence of radial symmetry. If it could be established that the carpoids were not antedated by radially organized echinoderms, and thus were not secondarily bilateral (like some forms discussed in the next chapter), then such "preradial" status would give them the utmost significance in echinoderm (and perhaps chordate) phylogeny.

Before discussing the peculiar functioning systems of echinoderms, it is perhaps worthwhile outlining the characteristic features of these five living classes, from whose common peculiarities we can then try to deduce something of an archetypic echinoderm. The true starfish, or Asteroidea, have usually five arms which are not clearly marked off from the central disc and these fleshy arms contain lobes of the alimentary canal and gonads and other internal structures. The ossicles of starfish are embedded in a tough but flexible skin. The Ophiuroidea, or brittle-stars, have arms which are sharply demarcated from the disc and do not contain lobes of the alimentary canal (Figure 11·1A). The ossicles, particularly in the arms, form an almost continuous articulated armor. The sea-urchins, or Echinoidea, many of which tend toward a spherical shape, lack arms and have the aboral side reduced (see Figures 10·1 and 11·3B). They have a continuous armor of fused ossicles, and also long movable spines which in many cases assist the tube-feet in locomotion. The so-called irregular echinoids show yet another return to bilateral symmetry, in this case associated with a burrowing life in the sea-bottom. The sausage-shaped sea-cucumbers and trepangs of the class Holothuroidea lack arms and have a body drawn out oral-aborally, and a leathery skin with few small scattered ossicles (Figure 10·1). The only surviving Pelmatozoa, the sea-lilies and feather-stars of the class Crinoidea, have a greatly reduced disc and the arms showing great development of branching (Figure 11·1B). They are stalked, or at least attached, for the whole or part of their life. Only a few living forms survive of this once enormous class and subphylum.

General Morphology and Functioning of Systems

The eleutherozoan echinoderms represent one of the few groups of invertebrate animals where setting up a functioning archetype contributes less to an understanding of the physiology of the group than does a thorough appreciation of actual functioning in a form such as the starfish, *Asterias*. This is so for two reasons. First, the archetypic echinoderm must have had a physiology close to that of present day crinoids, and totally unlike that in the other groups. Secondly, most physiological investigations have been carried out on starfish and sea-urchins, and the former group represent in many respects the least specialized (or most "archetypic") of the Eleutherozoa. For once, looking at a "type," the starfish, is a better way toward an understanding of physiology in the group.

Beginning with the skin and body wall, the diagnostic characteristics of the phylum are shown and, particularly on the oral surface of the starfish, the epithelium is ciliated. This epithelium is supported by the ossicles formed in the dermis, linked together by dermal connective tissues. The ossicles occasionally penetrate the surface as plates, or spines, and some are movable, especially along the sides of the ambulacral grooves. This results in a range of appearance in different starfish from a warty or tuberculate surface to a prickly "exoskeleton." Besides the calcareous protrusions, pockets of thin skin are developed. These papulae are mostly aboral and have a respiratory function, because within them there is a prolongation of the general visceral cavity (part of the coelom) and they thus form regions of increased surface area where respiration can go on. The similar protrusions within the ambulacral grooves (the podia) will be considered along with the water-vascular system. A fourth sort of skin appendage occurs on all surfaces except the ambulacral grooves. The pedicellariae are of various sorts (see Figure 11·2A) and their function appears to be protection of the papulae and the rest of the skin by capturing and holding small organisms. The pedicellariae in starfish are nonpoisonous though in sea-urchins they have associated poison glands. Once closed on an object, the pedicellariae remain so for some hours. To some extent the pedicellariae resemble the nematocysts of coelenterates in that they are apparently independent units which respond in a local reflex to a local stimulus. Again, like nematocysts, they may in some cases require a double stimulation, both mechanical and chemical.

The nervous system is always superficial, and this is even more obvious in starfish than in the other groups. The most obvious system—the oral or ectoneural nervous system—lies immediately under the thin ectodermal epithelium, with a circumoral ring and five obvious radial nerves running closely applied to the ambulacral grooves (Figure 10·2A). These are, however, only the most obvious elements in a general subepidermal plexus which covers the entire body surface. Further, the ectoneural nervous system is only one of three nervous systems in the starfish. The entoneural nervous system, which is more important in crinoids, runs along the ossicles at the sides of the ambulacral grooves and provides motor innervation to the muscles between the ossicles. It is also connected to a nerve net in the coelomic lining of the body wall. The third system, the hyponeural nervous system, has radial elements on the sides of the perihemal canals and is connected to the local motor ganglia of the tube-feet. Perhaps the easiest hypothesis about the archetypic organization of nervous structures in echinoderms would be that there were originally networks connected to six centralized systems, that is an oral and aboral manifestation of ectoneural, entoneural, and hyponeural systems. Under this hypothesis, the different vestiges of these systems which are important in the different groups of living echinoderms would be those which had functional significance appropriate to the group involved. Considering only the neural aspects of locomotion, the motor innervation to the muscles on the outside of the body

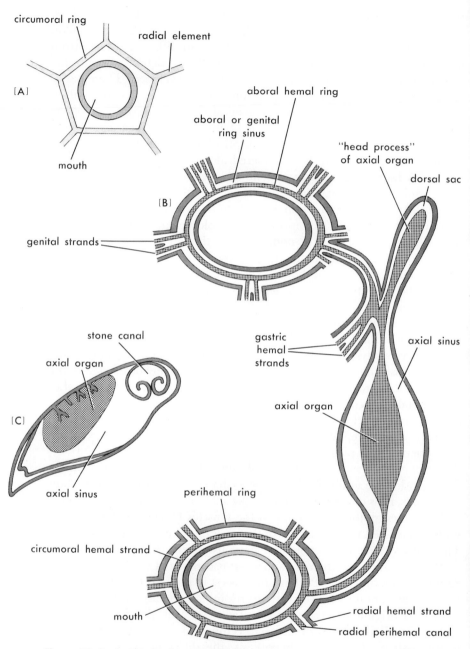

Figure 10·2. A: The basic anatomical plan for many organ systems in the echinoderms consists of a circumoral ring element and five radial elements. This diagram could serve for the central parts of the ectoneural nervous system, or of the oral hemal system, or of the water-vascular system. **B: Stylized perihemal coelomic and hemal systems** in a starfish. For further description, see text.

wall immediately below the epidermis would be important in one group while the motor innervation to the muscles between the ossicles would be important in another. Similarly, *oral* centralization might seem to be adaptively more significant in starfish while aboral centralization might be appropriate to crinoids. Such reduction to one appropriate "system" does in fact occur.

As regards the functioning of starfish, much of their behavior involves local reflexes, for example the local responses of tube-feet, of pedicellariae, and of movable ossicles. These local reactions are not dissimilar to those found in coelenterates such as sea-anemones (see BLI, pp. 15–16 and 48), and probably involve similar patterns of neural "spread," with decremental conduction and interneural facilitation both being involved. However, in starfish considerable "central" coordination is possible. The "righting" movements which are easily investigated by any student with a healthy starfish must be centrally controlled. The normal processes involved in feeding on bivalves also require overall coordination. The sense-organs are mostly diffuse and there are both tactile and chemoreceptors all over the surface. In most starfish the terminal tube-foot of each groove lacks a sucking disc and has an obvious pigment spot in its base which surrounds a group of ocelli.

Figure 10·3. Tube-feet on the oral side of a relatively primitive starfish, *Luidia ciliaris*. [Photo © Douglas P. Wilson.]

Echinoderm Guts

As with other organ systems, in the alimentary canal of echinoderms, in spite of nomenclature, we are considering organs which are certainly not homologous with those similarly named in other phyla. In typical starfish, the basic layout of the alimentary canal is that of a tube extending vertically from the mouth to the anus on the aboral, or upper, side. This last opening is almost nonfunctional. The minor elaborations of the system can be seen in Figure 10·4B. The stomach is completely eversible, both during feeding on small bivalves and such, and in extruding undigested waste. Most starfish are predatory carnivores, though a few species are known to be mucus-ciliary gatherers of particulate matter. Many textbooks suggest that toxins are secreted by the stomach wall during the feeding of forms like *Asterias* on small molluscs. It is most probable that only digestive enzymes are so secreted, though these can be used externally by folds of the everted stomach. Recent work has shown that in certain predaceous starfish, folds of the stomach wall are insinuated within the molluscan shell into its soft parts and release enzymes which begin the digestive process. This invasion of the prey by folds of the stomach wall can take place within the starfish in the case of small molluscs, or totally externally, so that the prey is digested outside the body of the starfish. Digestion is, of course, entirely extracellular (except possibly for some fatty substances). The ciliation of the central parts of the gut is all directed orally and is concerned not only with expulsion of undigested, particulate matter, but also with bringing to the folds of the stomach wall the powerful digestive enzymes which have been produced in the glandular pockets of the pyloric caeca. Little, if any, enzymes are produced by the gastric-wall epithelia themselves, and the functional significance of the intimate contact with prey tissues which seem necessary to successful digestion remains obscure. Digestion and absorption go on in both stomach and pyloric caeca. The latter are also concerned with enzyme secretion, with storage of nutritional reserves (masses of glycogen but also surprisingly high quantities of lipids), and, in some forms, with ciliary "sorting" of partly digested food material.

Something of the variation found in the guts of echinoderms is shown in Figure 10·4, the ophiuroid pattern being simplest as a blind sac without an anus and without any diverticula. The other three groups, crinoids, echinoids, and holothurians, have rather more elongate guts lacking diverticula. Recent work with radioactive tracers has shown that the main nutrient transport within starfish is through the coelomic fluid in the general visceral coelom. Such transport is probably also true of echinoids and holothurians; little is known about transport in ophiuroids or crinoids. It seems almost impossible in the latter group where the visceral coelom is open through a pore system to the sea. It is worth

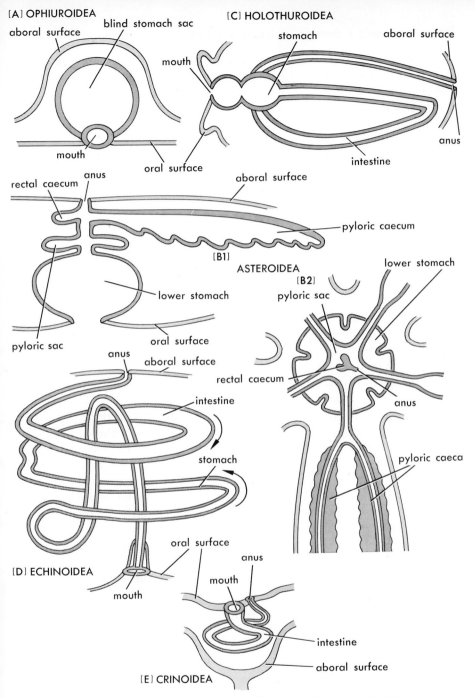

Figure 10·4. Gut patterns in echinoderms. A: The simple blind sac of brittle-stars. **B1** and **B2:** Lateral section and aboral view of the gut of starfish with its elaboration of diverticula. **C:** The elongate tubular gut of sea-cucumbers. **D:** The similarly elongate gut of regular sea-urchins with its peculiar apposed loops. **E:** The relatively minute tubular gut of crinoids, which terminates in an anus alongside the mouth.

mentioning here that the coelomic fluids of the visceral cavity also serve as the main respiratory transport system in the more massively bodied echinoderms: starfish, sea-urchins, and holothuroideans. There is no true circulation, but food components and oxygen are carried along diffusion gradients, aided to some extent by "tidal" movements of the coelomic fluid. It is likely that this method of nutrient and respiratory transport represents one of the physiologically limiting features of the echinoderm functional plan. Faster moving animals *require* a true circulatory system for these purposes.

Water-vascular System

As stressed earlier, many of the unique features of the functioning of echinoderms depend on the existence of the part of the coelom termed the water-vascular system (WVS). The system consists (see Figure 10·5) of a circumoral canal and five radial canals with paired side branches, each leading to an internal ampulla and its externally protruded tube-foot. Each ampulla connects between ossicles (starfish), or through pores in an ossicle of the ambulacral groove (sea-urchins), to a single tube-foot externally. There is also a single asymmetric canal leading from the oral ring toward the aboral surface. This "stone" canal opens into a small dorsal ampulla which connects through many pore canals to the external surface. The plate bearing these pore canals is termed the madreporite. As shown in Figure 10·5, there are also five pairs of small sacs inside the ring canal, termed Tiedemann's bodies, and some starfish, though not *Asterias,* have larger sacs (presumably for water storage), termed Polian vesicles, in approximately the same anatomical position. A slight water current can be observed to pass continuously in through the madreporite and this is thought to make good any water losses through damaged tube-feet to the exterior. There is, however, no clear experimental proof of this. In fact, it is claimed that some starfish with the madreporite surgically removed, or experimentally blocked, can live and move using the tube-feet as usual. Actually, each ampulla–tube-foot system seems to be functionally separated by valves from the rest of the WVS and experimental preparations can be made of a single ampulla with a single tube-foot which can be caused to continue to function. There is some variation in the microanatomy of tube-feet. In *Asterias* and other forms, there is a muscle-operated vacuum-cup sucker at the outer end. Other forms have long gland cells producing a sticky secretion, and still others have neither and possess tentacular or pointed tube-feet. All podia, however, have a similar basic microanatomy (which we know largely from the work of J. E. Smith), the walls of the podium, or tube-foot itself, having only longitudinal muscle (Figures 10·3 and 10·5B). Thus stimulation

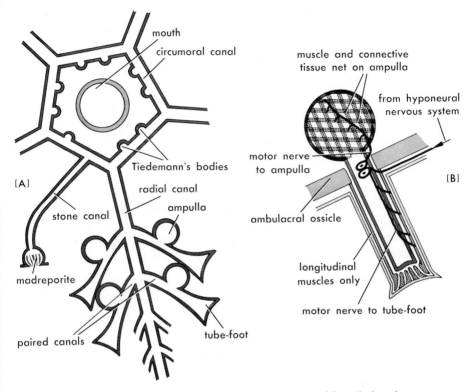

Figure 10·5. A: Stylized water-vascular system of a starfish. All the elements shown lie near the oral side except for the stone canal running to the aboral madreporite. **B: Stylized organization of a single tube-foot** with its ampulla. For account of general functioning and mechanics, see text.

of the tube-foot can bring about only its contraction or withdrawal. The ampulla is well-muscularized with a roughly rectangular meshwork of muscles and connective tissue surrounding the bag. Contraction of the ampullar muscles, with the valves to the rest of the WVS closed, causes extension and elongation of the tube-foot. We thus have a local effector unit working around a hydraulic skeleton. Local nervous arcs were demonstrated by J. E. Smith, and also connections of these to the hyponeural system through a ganglion cell near the neck of each ampulla, which acted as an individual motor center.

As well as starfish, echinoids and holothurians use their tube-feet in "stepping" locomotion. The forces applied to the substratum can involve both levering and pulling (or traction). In starfish, it seems that in normal horizontal movement the tube-feet are used mainly as levers, while in near-vertical movement, the traction method is used. Levering and postural bending is accomplished by contraction of the longitudinal muscles on one side of the tube-foot which work against the hydraulic pressure of the system, the pressure being maintained by partial contraction of the muscles of the ampulla. As already emphasized, it is difficult to imagine the myriad tiny locomotor units of a starfish creating any gross movement relative to the substratum unless they were supported by the unique skeleton of dermal ossicles, appropriately placed.

Another important function of the ampulla–tube-foot systems is to serve as respiratory organs. It has been demonstrated by A. Farmanfarmaian that the main sites of ingress of oxygen into the inside of sea-urchins is by way of exchange from the fluid within the ampulla to the perivisceral coelomic fluid. It is significant that in echinoids the ampulla is connected to the tube-foot by two canals passing through two pores in each ambulacral plate. In an expanded system a current of water passes continually out through one canal to the tube-foot and back to the ampulla via the other. It has recently been shown that even in starfish, similar ciliary currents cause a circulation of fluid to be maintained within each system, with water entering each tube-foot on the outside of the starfish arm and leaving it on the side toward the ambulacral axis. A number of workers believe that respiration must have been the archetypic function of the WVS and its podia. This attractive hypothesis suggests that these structures, unique to echinoderms, were first used by pelmatozoans for respiration and that they lay on either side of the food groove to derive advantage from the ciliary water currents along it. Later they became muscular tentacles which assisted in feeding, and still later became the organs of locomotion in the eleutherozoan echinoderms. Because of the mechanical importance of the recognizable ambulacral ossicles, stressed above, this is one of the few phylogenetic problems which may eventually be solved by paleontological work on the most ancient echinoderm fossils.

Hemal System and Gonads

As seen in starfish, the hemal system consists of strands of spongy tissue, which never form distinct vessels with walls, each one enclosed within a closed coelomic channel or perihemal space. The main layout consists (as usual) of a circumoral ring and five radial hemal strands on the oral side (Figure 10·2B). There is a single axial organ enclosed within the axial canal which runs in close association with the stone canal of the WVS to the aboral surface where there is an aboral, or genital, ring sinus containing the aboral hemal ring and its genital branches. There is some slight contractility of the axial sinus, particularly where it surrounds the head process of the axial organ near the aboral side. A few years ago this was claimed to constitute, in echinoids, a heart within a circulatory system. There is no evidence whatsoever of true circulation or even of a connected system of lumina within the axial complex, or anywhere else in the system of hemal, spongy tissues. It may be that the contractility is a functional vestige of the contractile structures, long known in the larval stages of echinoderms. More recently, it has been demonstrated that surgical removal of the axial complex does not affect the respiratory rates in some echinoids, thus disposing of one aspect of circulatory function. Further, recent work by Norman Millott has demonstrated that the functional significance of the axial organ itself lies in its capacity to respond to invasive particles or organisms within the visceral coelom of the echinoderm. By producing large numbers of amoebocytes in a sort of "immune response," it plays an important part in the defense of the echinoderm against both mechanical injury and disease organisms.

Another great importance of the hemal system lies in its developmental connection with the gonads of echinoderms. In starfish, the genital system consists typically of ten gonads, two in each arm. In young stages, each gonad is enclosed in a genital sac which is an outgrowth of the genital, or aboral, sinus of the perihemal-coelomic system. The sexes are normally separate and usually each gonad has a separate, small gonopore to the exterior near the base of each arm. In many cases, a stimulus to spawning is provided by the presence in the water of gametes of the other sex. In many echinoderms, fertilization is external and some aspects of the pelagic larval development with its metamorphoses will be discussed in the next chapter.

Echinodermata II: Variant Functional Patterns

JUST AS THE DIAGNOSTIC anatomical features of the echinoderms as a whole have a basis in a unique pattern of functioning—involving the dermal ossicles as skeleton and the hydraulically operated podia as effectors in a pentaradiate organization—so each of the five major living classes has a rather clearly defined structural pattern of functional significance. Within each class, more extensive homologous features, both structural and functional, can be discerned, and there is also a "way of life" or general pattern of ecology which is common to the members of each major subdivision.

Brittle-stars

The significantly different features of form and function in the Ophiuroidea, or brittle-stars, seem to spring from the use of the whole arms as the locomotor organs with a corresponding reduction in the importance of tube-feet. The long, slender, muscular arms imply not only peculiarities of the ossicles as a skeleton for these muscles, but also that there are no lobes of the gut or gonads outside the central disc. Further, the podia are without suckers, are often reduced, and they are protracted by a system of vesicles and head-bulbs which is probably less efficient than the ampulla–tube-foot system already described (which is possessed by all the remaining motile echinoderms).

Brittle-stars are surprisingly abundant animals. Their powers of homeostasis being slight, they do not occur to any extent between tidemarks and are thus rarely seen by shore-collecting students. However, throughout the oceans of the world, on shallower and deeper bottoms, brittle-stars are among the most abundant, successful animals (Figure

Figure 11·1. Brittle-stars and feather-stars. A: *Ophiocomina nigra,* the black brittle-star, and *Ophiothrix fragilis,* the European "common" brittle-star. **B:** *Antedon bifida,* specimens of the rosy feather-star, with at the bottom a small regular sea-urchin covering itself with shells. [Photos © Douglas P. Wilson.]

11·1). For a time after the Second World War, when techniques of submarine photography and underwater television were being developed, it seemed that almost every photograph or television probe of the bottom showed heaps of brittle-stars or regularly spaced patterns of them like carpets. Most ordinary brittle-stars can be described as carnivore-scavengers. They are predatory upon all suitable small organisms, but, on the soft substrata where they abound, spend much of the time sorting through the detritus and selectively picking out organic material for ingestion. The aberrant group of Euryalids (Figure 11·3A), and perhaps some other brittle-stars, feed by waving their arms in the water and catching planktonic organisms—mainly small crustacea.

In ophiuroids, the axis and much of the arm is occupied by a series of massive ossicles which have been termed vertebrae (see Figure 11·2B), to which are attached four series of longitudinal muscles. These can carry out different movements depending on the articular surfaces of the ossicles, but in most ophiuroids simultaneous contraction of all four muscles between two adjacent vertebrae will bring about autotomy or breaking off of the arm. Most brittle-stars use this as an escape reaction. All have great capacity for regenerating not only arms but parts of the disc and internal organs.

Living species of the class fall into two very unequal groups. The Ophiurae, or typical brittle-stars, have vertical articular surfaces in contact between the vertebrae, and this allows mainly movements in the horizontal plane. There are about one thousand nine hundred species in this group, which makes it the most numerous order of the phylum Echinodermata. The other order—the Euryalae or basket-stars—consists of forms where the arms can move vertically as well as horizontally and can twine around objects, as a result of the articular faces having "hour-glass" surfaces (see Figure 11·2D). Typical basket-stars like *Gorgonocephalus* (see Figure 11·3A) have branched arms although some allied genera, including *Astrotoma,* have simple arms. There are only a few genera and species of euryalids and they live mostly on hard substrata at considerable depths.

The other anatomical modifications of the arm in ophiuroids (see Figure 11·2B) are best understood if it is realized that the vertebrae are the ambulacral ossicles from the floor of that groove moved into the anterior of the arm. The ambulacral groove has become closed, that is the radial nerve is no longer at the surface but lies on the inner surface of a new canal formed by the closure of the groove—the epineural canal—and the general visceral coelom is squashed up aboral to the vertebrae. This was probably the way the ophiuroid type of arm evolved, and it is certainly the way in which the structures develop embryonically. During development the open ambulacral groove is replaced, and some ophiuroid vertebrae show throughout life the "suture" line of the fusion of the two ambulacral ossicles which gave rise to it. Above the epineural canal and the radial nerve lies the radial perihemal canal with a hemal strand and hyponeural nerves. In brittle-stars the relatively thin ribbon of the hyponeural system lies just above, and closely applied to, the thicker strand of the ectoneural radial nerve. It can be demonstrated that the hyponeural is pure motor nerve while

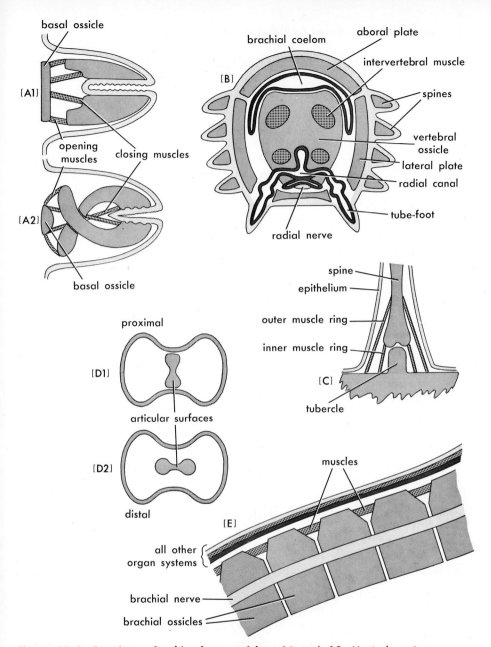

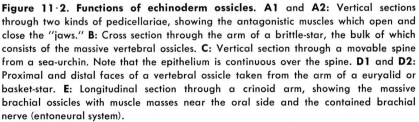

Figure 11·2. Functions of echinoderm ossicles. A1 and **A2:** Vertical sections through two kinds of pedicellariae, showing the antagonistic muscles which open and close the "jaws." **B:** Cross section through the arm of a brittle-star, the bulk of which consists of the massive vertebral ossicles. **C:** Vertical section through a movable spine from a sea-urchin. Note that the epithelium is continuous over the spine. **D1** and **D2:** Proximal and distal faces of a vertebral ossicle taken from the arm of a euryalid or basket-star. **E:** Longitudinal section through a crinoid arm, showing the massive brachial ossicles with muscle masses near the oral side and the contained brachial nerve (entoneural system).

the ectoneural involves both sensory and motor elements. Corresponding to each vertebral ossicle is a ganglion on the compound radial nerve, but there is a much greater degree of central nervous control than in the asteroids. For example, if the circumoral nerve ring is cut in two interradii, so as to isolate an arm, that arm cannot show any coordinated movement and is simply towed behind limply as the brittle-star moves along. In view of the relatively simple arrangement of the arm muscles (which in normal locomotion involve the muscles of one side as antagonists of the muscles of the other), it is significant that, in addition to autotomy, gross overstimulation of brittle-stars causes them to freeze or stiffen. All these escape reactions—autotomy, freezing, etc.—involve simultaneous contraction of all the intervertebral muscles.

Another major modification of brittle-star anatomy is that the aboral surface grows down interradially and around on to the undersurface. Thus the madreporite is moved to the oral surface at one side of the mouth. On either side of each arm as it nears the mouth are slits in this underturned aboral surface and these ten slits lead into the genitorespiratory bursae which are lined with respiratory tissue and into which the ten gonads discharge. As already stated (see Figure 10·4A) the gut is extremely simple. A short wide buccal region with teeth leads through an even shorter oesophagus to the blind sac of the stomach. There are no lobes, no intestine—not even an anus. Five independently moving jaws with teeth surround the mouth, and the first pair of podia on each arm form sensory buccal tentacles. Obviously these tentacles and delicately moving jaws are important in the selection of valuable food material from the detritus of the sea bottom. The tiny gut with its blind stomach, simply occupying all the visceral cavity not occupied by other organs, is obviously unsuitable for wholesale ingestion of the entire substratum regardless of nutritive value, as is done by holothurians and certain echinoids.

Sea-urchins

The functional pattern of the class Echinoidea is based on the fact that the ossicles of the endoskeleton form an external armor, or test, of closely fitted plates arranged around a globose body, and this allows movable spines operated by dermal muscles to be used along with the podia in locomotion. Characteristically they have a reduced aboral surface and an extended oral surface (see Figure 10·1), and the test consists of plates arranged in twenty meridional rows, ten being ambulacral with pores for the tube-feet, and ten being interambulacral and usually larger (Figure 11·3B). The more rigid skeleton thus formed provides a basis for the action of the movable spines. Most regular sea-urchins are macrophagous herbivores, feeding on attached algae or

Figure 11·3. A: Oral view of a basket-star, the euryalid ophiuroid *Gorgonocephalus*.
B: Aboral view of the test of a regular sea-urchin, *Arbacia*, showing the alternation
of the ambulacral (darker) and interambulacral double rows of plates arranged
meridionally, and the periproct (anal) area surrounded by five large "basal" plates
at the ends of the interambulacral rows, four of which show clearly the genital pores,
while the fifth (at the top) is the madreporite, with numerous minute pores in addition.
[Photos by the author.]

other large plants for preference, but they seem to be potentially omnivorous. Their preference for grazing on living larger algae, means that
they are typically a group of the immediate and illuminated sublittoral.
In life, a ciliated epithelium, which is often very thin, covers all the
test, spines, etc. The ossicles are less fused around the two poles. This
results in two circular leathery areas with separate ossicles termed the
peristome and periproct. Besides the pores of the tube-feet, the ossicles
bear patterns of bosses to which are attached both spines, which are
often of two sizes, and pedicellariae which can be of four kinds. These
pedicellariae often have a three-jawed arrangement of movable ossicles and some have a poison gland in each jaw. Spines each have a
double muscular system with the outer muscles causing the movements
and the inner serving as catch muscles (Figure 11·2C). Thus the inner
ring of tonic muscles holds the spine erect on the tubercle of the test
ossicle, and the outer ring of phasic muscles is responsible for pointing
the spine in any particular direction. Although most knowledge of the
nerve supply to these muscles is inferential, it is obvious that the spines
are capable of considerable coordination in their more delicate movements. Apart from locomotion, the work of Norman Millott and others
has revealed elaborate patterns of photic responses, which are used by
some sea-urchins (notably the black *Diadema* of coral reefs) for ag-
gressive defense, and by others for constructing concealing camouflage.

The water-vascular system is similar to that in starfish with a circumoral ring with five Polian vesicles and five radial canals running meridionally, supplying the paired vessels to the ampullae. As already mentioned, each ampulla has two canals running through two pores in the ambulacral ossicle to each tube-foot and this is thought to have considerable respiratory significance since the water current circulates in each unit under the influence of the cilia lining the lumen of the tube-foot and ampulla. The hemal system and axial gland are again similar to starfish, and in spite of recent claims, there is no evidence of true circulation in the hemal system. The five gonads are suspended by mesenterial strands in the interambulacral regions. In the adult, there is no trace of the genital stolon ring, though it is present as a coelomic derivative during early adult development. The main nervous system is ectoneural with circumoral and radial elements, but this is connected to a subepidermal plexus running all over the surface outside the test. Much experimental work on the movements of spines has involved experimental cutting of this plexus and of the radial nerves. In crude summary of this work, the spines point toward the source of stimulus in detached pieces of sea-urchin body wall, and it can be demonstrated that the stimulus is transmitted through tissues on the outside of the test.

As already noted, the alimentary canal is more complex and elongated (see Figure 10·4D) and the mouth is surrounded by a complicated skeletal framework which has been termed Aristotle's lantern. This supports five chisel teeth, each attached to a complicated system of muscles and coelomic spaces. This complicated structure (the lantern itself consists of forty skeletal pieces) allows regular sea-urchins to protrude and turn the whole lantern, to open and close teeth, and to move individual teeth in all appropriate directions. The teeth themselves are secreted continually from sacs at their internal ends and their outer ends are hardest.

The urchins which are termed irregular, and which belong to at least two distinct orders, show a secondary bilateral symmetry. The anus and the periproct area is no longer apical but lies in the posterior interradius and the aboral parts of the ambulacra are expanded like petals and bear flattened respiratory tube-feet. These forms are all sand-burrowing and this may partially explain their development of a real anterioposterior axis. Forms like the heart-urchin *Echinocardium* have the mouth placed anteriorly and lack the jaw apparatus of regular urchins. The sand-dollars (including *Clypeaster* and *Mellita*) are much more flattened but have the mouth placed centrally below and retain the lantern structures. Zoologists have long used an oversimplified classification of the urchins: into two orders of regular urchins and two orders of irregular urchins. However, the Echinoidea constitute a group where neontologists can no longer afford to ignore the fossil record, here so abundant and so well studied. H. Barraclough Fell (one of the rare individuals who has worked on both living and fossil forms) has proposed a classification of the Echinoidea which involves nineteen orders, fifteen of which have recent representatives.

Sea-cucumbers

The class Holothuroidea are characterized by two features which have allowed them to become the only really successful burrowing forms among the echinoderms. First, they have a muscular body wall with very small, widely separated ossicles (without spines or pedicellariae externally). Secondly, they are sausage-shaped with an elongation of the disc part of a starfish body in an oral/aboral direction but without any arms. The new wormlike patterns of locomotory musculature have allowed degeneration of a major archetypic feature: the intimate association of the podial system with dermal ossicles is no longer necessary. In some—including many deep water forms—there is a secondary bilateral symmetry which arises from one side of the body becoming a sole on the ground. When this happens, usually three radii of the ambulacra have more functional tube-feet and become the ventral sole. The tentacles around the mouth—which are modified tube-feet—may be peltate (so-called shield-shaped), or pinnate with a limited number of lateral branches, or very much branched or dendritic. Their retraction may involve a hydraulic system with large ampullae, as in the peltate ones, or retractor muscles as in the dendritic ones, or either or both, as in the forms with pinnate tentacles. Certain sea-cucumbers, including *Cucumaria* and its allies, are plankton-feeders, catching minute organisms on the sticky mucus secreted from the expanded tentacles and rhythmically wiping the loaded tentacles through the mouth in turn. Almost all other holothurians ingest masses of the substratum, like certain worms, and extract organic food material from it (Figure 10·4C).

In most, there are two branched long ducts, the so-called respiratory trees, running in from the cloaca. Muscular pumping action of the cloaca results in exchange of seawater into the tree system and there is an oxygen exchange with the fluid of the visceral cavity. In *Holothuria* and its allies, the more posterior branches of the system are termed Cuvierian organs and are extrusible for defense as sticky threads. They always wriggle actively after extrusion, and in some cases may be toxic. Under gross stimulation, many sea-cucumbers show an escape reaction involving expulsion of all the viscera through the cloaca. In many, regeneration of the viscera can be accomplished in a few weeks.

Most of the water-vascular system is anatomically similar to those already discussed. In some sea-cucumbers there is a secondary multiplication of the stone canals, and in most the madreporites are internal, opening within the visceral cavity in adults. It is clear in development that each tentacle is an enormously enlarged podium, or tube-foot, and in some sea-cucumbers it is extruded by an ampullar hydraulic system. The hemal system is more greatly developed than in other echinoderms though the axial organ and its sinus are reduced. There are two large lacunae with elaborate branches lying alongside the S-shaped alimentary canal, and the subdivided branches of these become involved with the respiratory trees in some sea-cucumbers. Once again, there is no real evidence of a circulatory system, although there are characteristically enormous numbers of amoebocytes associated with this hemal system.

Normal and Aberrant Starfish

As already discussed, starfish, or sea-stars, are normally regarded as predaceous carnivores, feeding especially on bivalve and other molluscs. Such prey if small is ingested, but if large is consumed extraorally. This involves extrusion of the stomach lobes as already discussed, but never involves toxins, although these are invoked by most textbook writers. The forces applied to the bivalve shell by the numerous tube-feet have been the subject of recent investigations, and forces of the order of five thousand grams can be applied by starfish such as *Asterias* and *Pisaster*. These forces are applied by a contraction of the longitudinal muscles in the tube-feet as is used in "pulling" locomotion.

A number of "aberrant" starfish are apparently ciliary–mucus-feeders. Over fifty years ago, J. F. Gemmill in Scotland demonstrated ciliary feeding in *Porania*. This was largely forgotten, until in recent years John M. Anderson, and subsequently B. N. Rasmussen, have clearly demonstrated that *Porania,* species of *Henricia,* and other forms are ciliary particle-feeders. The ciliation in the pyloric stomach and caecum of *Henricia* (as described by Anderson) resembles nothing so much as the gut-sorting mechanisms in particle-feeding molluscs.

With regard to the phylogeny of feeding patterns within the echinoderms as a whole, it would be satisfactory to assume that the "normal" macrophagous starfish were derived from ciliary-feeding, more primitive, forms. Unfortunately, this seems not to be so, as the most archaic living starfish are the relatives of the genus *Luidia,* species of which have been observed to be voracious carnivores.

This group, with one living representative genus, *Platasterias*, has a fundamental pattern of arm ossicles which is intermediate between that typical crinoids and that of the starfish. Somasteroids resemble flattened starfish, but have a blind gut with no anus, and small tube-feet without suckers. The significance of this stem group has only recently been elucidated by H. Barraclough Fell.

It seems that *Platasterias*, at least, has a feeding mechanism resembling those found in crinoids, and like them is a microphagous gatherer. Similar feeding habits probably sustained the more numerous fossil somasteroids, and perhaps also the entirely extinct blastoids and cystoids.

Feather-stars and Sea-lilies

Obviously, the living species of the class Crinoidea represent the survivors of the most archaic stocks of echinoderms. As their position in this discussion, and the inversion of Table 10·1, are meant to indicate, unfortunately less is known about their physiology than that of all the other groups. This is because most of them survive in relatively deep water, and they are very difficult to maintain alive in aquaria.

Of the group, more than five thousand species have been described as fossils, but only about six hundred and twenty living forms of which eighty are sea-lilies (i.e., are stalked and permanently sessile). Their near-limitation to the deeper parts of the oceans suggests a group going to extinction, but they were once abundant as the thick beds of certain Carboniferous limestones, formed from crinoid ossicles, indicate. *Antedon* is typical of the more numerous comatulids, which can swim and attach only temporarily (Figure 11·1B).

The body wall is mainly ossicles and the epidermis is not continuous over these. The ten arms are made up of the brachial ossicles (see Figure 11·2E) with muscle masses near the oral side and a big brachial nerve running through them. The podia which are directly connected to the radial elements of the water-vascular system without ampullae, form a double row of tentacles on the sides of the ciliated grooves which are the feeding mechanism. These split radial canals lead to a circumoral canal, but there is no external madreporite: openings from numerous stone canals are within the perivisceral cavity. The hemal system is reduced in the adult but there is a ring (or genital rachis) from which genital strands run down each arm and are enlarged into gonads in the pinnules. There is an ectoneural nervous system (as in starfish) with radial nerves lying immediately below the ciliated grooves of the food-collecting mechanism. Much more important for movements is the aboral entoneural nervous system which forms a cup-shaped mass below the aboral calyx, and has branches from the ring nerve around it into the ten arms. These ten brachial nerves supply the motor fibers to the muscles between the ossicles, as can be demonstrated by destructive experiments. The alimentary canal runs from a central mouth through a short oesophagus into the stomach, once round the disc, from which coil, in *Antedon,* arise peculiar small diverticula of unknown function. The gut runs then through the short rectum to an anus placed interradially on the oral side. Crinoids apparently live on a mixed diet of detritus, microorganisms, and plankton. The larger elements appear to be trapped by an interlocking net formed of the pinnular tentacles, while a mucous secretion, produced by the small pointed podia, traps smaller food particles which are then transported by the ciliated ambulacral grooves which run together into the mouth.

Echinoderm Larvae

No other group of animals has such complicated metamorphoses in the course of development. A few echinoderms brood their young but, in many, fertilization is external and the zygote develops in the sea. The first cleavages are relatively regular and indeterminate, thus providing the famous material used by the earliest experimental embryologists. The embryo remains holoblastic to the blastula stage. Gastrulation is followed by various patterns of larval development, which always involve bilaterally symmetrical stages for varying periods. This is then followed by a metamorphosis during which the left anterior part of the larva gives rise to most of the radially symmetrical adult structures. The major larval types are illustrated in Figures 11·4 and 11·5: the bipinnaria of starfish (Figure 11·4A) resembling the auricularia of sea-cucumbers (Figure 11·5B), and the ophiopluteus of brittle-stars (Figure 11·5A) resembling the echinopluteus of sea-urchins.

In the echinoderm gastrula, a wide blastocoel, or segmentation cavity, separates the enteron wall from the ectoderm. The blastopore becomes the anus, and a mouth is formed by a breakthrough of the stomodeum in a "typical" deuterostomous fashion. Many texts state that, at this stage, all echinoderm embryos develop three pairs of segmentally arranged coelomic sacs from enterocoelic vesicles. This idealized development has been elevated to give a hypothetical ancestral form for the echinoderms, the dipleurula. In its simplest—and perhaps its only useful—manifestation, this theoretical organism is regarded merely as an idealized larva combining many of the features common to all eleutherozoan larvae. Thus it shows three pairs of coelomic

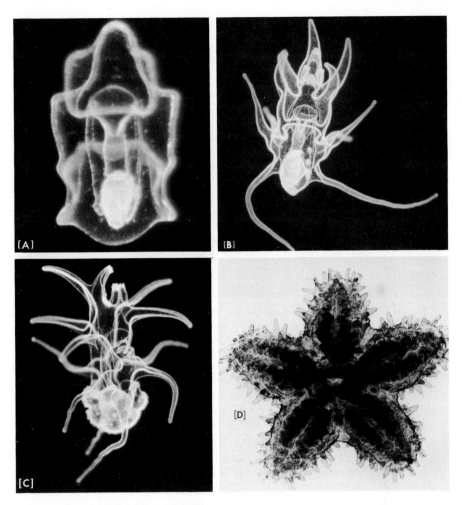

Figure 11·4. Starfish larval development: stages of *Asterias*. **A:** Bipinnaria larva, a ventral view showing the complete bilateral symmetry. **B:** Early brachiolaria larva in ventral view. **C:** Later brachiolaria larva in lateral view, showing the radially symmetrical rudiment of the starfish body beginning to form. **D:** Young starfish after metamorphosis. [**A, B, C:** Photos © Douglas P. Wilson. **D:** Photo by Marvin E. Snow of a specimen prepared by Dr. Albert J. Burky.]

spaces, but omits both the specializations of the pluteus larval types and those of the auricularia-bipinnaria group. In more extreme—and largely indefensible—hypotheses, the dipleurula is enlarged, given gonads, complex sense-organs, and a benthic crawling habit, and is set up as a hypothetical preradial echinoderm ancestor. When larval types are surveyed throughout the Echinodermata, there are several examples of close larval resemblances between unrelated species (reflecting

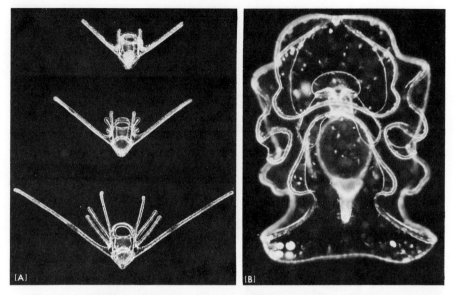

Figure 11·5. More echinoderm larvae. A: Three successive ophiopluteus stages of *Ophiothrix fragilis,* in ventral view. The elongate ciliated *larval* arms are bilaterally symmetrical and completely lost at metamorphosis to the adult brittle-star. **B:** The auricularia larva of the sea-cucumber, *Labidoplax digitata,* in ventral view. Compare with the starfish bipinnaria larva of Figure 11·4A, and the hemichordate tornaria larva of Figure 12·2 and, for further discussion, see text. [Photos © Douglas P. Wilson.]

evolutionary convergence), while other closely related forms exhibit major differences of larval types (divergence). Detailed phylogenetic relationships cannot be extrapolated from larval similarities in echinoderms. In broader questions of lineage, larval features can perhaps provide corroborative evidence but cannot be accepted as the sole proof of relationships. Actually, a twofold and threefold subdivision of the larval coelom does occur in several different groups, but not with the regularity that certain books suggest. There is, however, considerable constancy (except in the case of the axial organ) in the derivation of the adult coelomic structures from the larval segmental sacs, particularly those of the left side. In a few sea-stars, there *does* occur the development of the coelom which is claimed as archetypic, and of the three pairs of coelomic sacs in the larva: the posterior spaces give rise to the visceral coelom and perihemal coelom, the left anterior gives rise to the axial sinus and stone canal, and the left intermediate coelom of the larva gives rise to the rest of the water-vascular system of the adult.

At the blastula stage of development, cilia cover the surface of the embryo evenly, but as the bilaterally symmetrical larva develops further, they become restricted to characteristic bands. In crinoids, the barrel-shaped larva has five ring bands of strong cilia surrounding the body. In the other four groups, the ventral surface of the larva becomes concave and this depression is outlined by the major band of cilia which is the main organ both of locomotion and of feeding. This grows much faster than the rest of the surface ectoderm, and becomes thrown into folds on special projections. These are the larval lobes, or arms, which are always arranged in bilateral symmetry. (See Figures 11·4A and 11·5A; these lobes bear no relation to the adult arms.) In the sea-cucumbers, the elongate larva has short blunt lobes. In this auricularia larva, the ciliary band forms a figure-of-eight before dividing into a smaller preoral ciliary band and a larger band posteriorly (Figure 11·5B). In Asteroidea, there is a similar larva, with somewhat greater lobe development, termed a bipinnaria (Figure 11·4A), which shows complete separation of the preoral band from the rest of the ciliation. In brittle-stars and sea-urchins, there is an enormous development of long temporary arms with internal calcareous skeletal supports. Both show relatively slight preoral development and the ophiopluteus (Figure 11·5A) and echinopluteus show remarkable resemblances to each other. In the later development of some sea-cucumbers, the auricularia becomes a pupa larva with five barrel rings like a crinoid larva. Both crinoids and asteroids become attached preorally before metamorphosis. In crinoids a new mouth is developed posteriorly and there is always a miniature stalked sea-lily stage, even if the adults (like *Antedon*) are free feather-stars. In the other four major classes, a new mouth develops on the left side of the larva and the radially symmetrical organs form around it. For example, in the brachiolaria, the starfish-rudiment forms inside, on the left of the gut (see Figure 11·4C).

Phyletic Thoughts

In spite of our much more sophisticated knowledge of the fossil groups of echinoderms, and the realization that the Eleutherozoa form an unnatural assemblage, it still seems justifiable to separate the crinoids and the several fossil groups allied to them in a division which may still be called the Pelmatozoa (or else the Crinozoa). All these forms seem to be clearly distinct, probably more ancient, and are undoubtedly not only basically sessile in habit but almost certainly microphagous with ciliary- or mucus-feeding on the outstretched arms. Among the mobile echinoderms, or Eleutherozoa, the Somasteroidea are clearly closest to the crinoids, but the Asteroidea also share certain possibly primitive characters. These include attachment at metamorphosis, and the exposed ectoneural nervous system. If we accept the customary classification as reflecting natural relationships, then we emphasize the features which link the echinoids and ophiuroids, including the larval similarities, and the common features of anatomy involving the canals associated with the ambulacral groove and its ossicles. If we accept the relationship suggested by Fell, which stresses the affinities between holothuroideans and echinoids on the one hand, and ophiuroids and asteroids on the other, then we give greatest phyletic significance to his distinction between the meridional gradients of the former and the radially divergent growth gradients of the latter. Whichever interpretation of the interrelationships of the living groups of echinoderms is more correct (and it may be very difficult to settle this), considerable evolutionary convergence is implied. In the one case, the resemblances between the ophiopluteus and echinopluteus must be due to convergent larval evolution, and the anatomical similarities of the ambulacra result from similar processes. Alternatively, the two dominant gradient patterns established by the rudiments of the water-vascular system, which subsequently control the development of the dermal ossicles and all other structures, must each have arisen twice (again in a startling example of convergent evolution).

However difficult it is to sort out the interrelationships within the echinoderms, there are some important phyletic conclusions regarding the group as a whole which—although hypotheses like all phylogeny—are accepted by the majority of zoologists. The following seem the most significant points. To a great extent, radial symmetry with the mouth up is functionally useful to sessile animals which have the same relationship to their environment on all sides. On the other hand, bilateral symmetry with a front and rear end, upper and lower surfaces, left and right pairs of effector organs, is functionally suited to the needs of a traveling animal. Apart from the echinoderms, the only other group of metazoan animals in which radial symmetry can be the rule is the coelenterates, many of which are fixed. Now, in more archaic living echinoderms there is attachment to a hard substratum at metamorphosis. Further, in addition to the six groups with living representatives, there are at least five other major groups, at least as distinctive, found only as fossils and now completely extinct. It is probable that many of these were sessile organisms, and it is also most probable that the earliest echinoderms were all fixed. Thus it seems likely that all echinoderms were at one time fixed and those now mobile retain the radial symmetry of these ancestors. If we go a little beyond this hypothesis, the ontogeny of an irregular echinoid, for example, involves a radially symmertical gastrula giving rise to a bilaterally symmetrical larva, giving rise to a radially symmetrical preadult, giving rise to a secondarily bilaterally symmetrical adult. Apart from the secondary return to bilateral symmetry, which is special to the case of irregular echinoids, this ontogeny suggests a phylogeny involving a bilaterally symmetrical ancestor before the radially symmetrical one which is also common to all groups.

It is worth stating first that the concept of the dipleurula larva is grossly overemphasized in most discussions, and secondly that the metamorphosis of typical living echinoderms is so complete that it is very difficult to speculate on presettled ancestors. Once again, it is important to note that certain characters common to several larval types need not reflect ancestral characters, but can simply be characters required by all larvae with that particular mode of larval life. Further, as several authors have noted, it is unlikely that a fossil, common preradial ancestor will ever be discovered since it is likely that only after settlement and the acquisition of radial symmetry that endoskeletal ossicles developed. However, the enigmatic carpoids could represent this unlikely fossil stock. As already stressed here, only when the group became echinoderms, by the development of such ossicles, was the further development of the characteristic hydrocoel, or water-vascular system, with its multipurpose podia, functionally likely. As regards the phylogeny of this diagnostic feature of the group, in all living forms this unique division of the coelom is found, and it is concerned with extending the podia hydraulically out into the environment. Whether these podia were first most important as respiratory, excretory, locomotory, or sensory structures is an unresolved question. It seems clear that very early in the history of the echinoderms, they were involved in the food-collecting apparatus of the sessile forms, both as mobile food-collecting tentacles and as the sources of a mucous net. They function thus to this day in the crinoids and perhaps elsewhere. Almost certainly, the evolution of internal ampullae isolated from the rest of the system by valves, the development of strong postural musculature, and the development of suctorial discs were later developments in the evolution of these organs. Such true tube-feet allowed the efficient locomotion of starfish and the forceful opening of bivalves for food. The fantastic elaborations of the feeding tentacles of sea-cucumbers, and the funnel-building tube-feet of irregular echinoids, were obviously still later elaborations.

CHAPTER 12

Invertebrate Chordates I: Diagnostic Physiology

DISCUSSION OF VERTEBRATES—animals with a jointed axial skeleton of vertebrae—lies outwith the scope of this book. Vertebrate animals—fish, amphibians, reptiles, birds, and mammals—make up over 95 per cent of the species of the phylum Chordata. The remaining forms which fall within the diagnostic definition of the chordates are usually described, along with their allies in the phylum Hemichordata, as "the primitive chordates" or "invertebrate chordates." Some of the great biological interest in these forms lies in the hypothesis that they represent prevertebrate protochordates, examination of whose physiology should give us some ideas of functioning in the stocks of animals ancestral to the vertebrates. In fact, understanding of vertebrate physiology and evolution *has* profited greatly from recent detailed work on the physiology and ecology of primitive chordates.

The Protochordata do not form a natural assemblage, and in the classification used here fall into one phylum, Hemichordata, and two distinct subphyla of the phylum Chordata. These are the subphylum Cephalochordata or Acrania, for *Amphioxus* and its allies, and the subphylum Urochordata, or Tunicata, for the more numerous sea-squirts. (There is, of course, a third subphylum in the phylum Chordata: the Craniata or Vertebrata for the approximately forty-three thousand species of backboned animals.)

Functional Significance Behind the Diagnostic Features

The diagnosis of the Chordata is relatively clear-cut, and can be stated —in anatomical and embryological terms—in about five main items. First, the notochord, a dorsal, axial skeleton, is present at some stage in the life-cycle. Secondly, pharyngeal gill-clefts open to the exterior from the gut and, having certain consistent features of anatomy and development, occur at some stage. Thirdly, there is a dorsal, central nervous system derived from the surface epithelium of the embryo, and usually in the form of a dorsal hollow nerve cord. Fourthly, the coelom is said to be enterocoelous in origin, and primitively divided into three divisions. Fifthly, there can be—as in no other invertebrates—a post-anal propulsive tail, that is, blocks of muscle around an axial skeleton posterior to the anus. The other features which could be used, including those of muscle physiology and ciliary control, do represent what are probably significant differences in the chordate stocks, but are not sufficiently consistent to be involved in a precise diagnosis. Of the five features of true chordates mentioned above, three (the second, third, and fourth) are part of the diagnosis of the phylum Hemichordata.

The "unknowable" ancestor of the chordates must have been a working, efficient machine involving most of these structural features. Thus in setting up a working archetype for the chordates—one in which the concert of organs and functions can operate as a whole—it is important to survey the functional significance of these diagnostic anatomical features in the more primitive living forms. While surveying the nature and extent of each feature in the main primitive chordate groups, it is worth also examining their physiological significance, especially in regard to such mechanical aspects as feeding and locomotion.

The notochord is an axial rod of peculiar vacuolated cells, which runs from end to end of the body in the Cephalochordata where it is present throughout life, and is the longitudinal skeleton of the larval tail in the Urochordata. It is no longer claimed as a feature of the Hemichordata, although some cells involved in the proboscis neck may be developmentally homologous with it. In the subphylum Craniata it is, of course, an embryonic feature replaced in the adult by the jointed, bony or cartilaginous vertebral column. Functionally, the notochord is a skeletal structure, with the mechanical properties of being laterally flexible while being incompressible from end to end. These mechanical properties arise from the turgidity of the large vacuoles of the cells. In the functional integration of the archetypic chordate, this type of skeleton must always have been associated with laterally placed blocks of muscles which could act as antagonists to each other, thus producing the characteristic lateral swimming flexures of the group. Propulsion by a laterally flexing tail with an internal skeleton is unique to the chordates and structurally and functionally homologous from sea-squirt larvae to newts.

The gill-clefts, which are openings from the pharyngeal part of the gut to the exterior, are present in all groups of chordates, though only in the embryos of the higher vertebrates. There is a tendency for their numbers to be reduced in the higher craniates, while they tend to be multiplied in the cephalochordates and urochordates. In all the proto-chordate groups, they multiply by splitting by tongue bars (see Figure

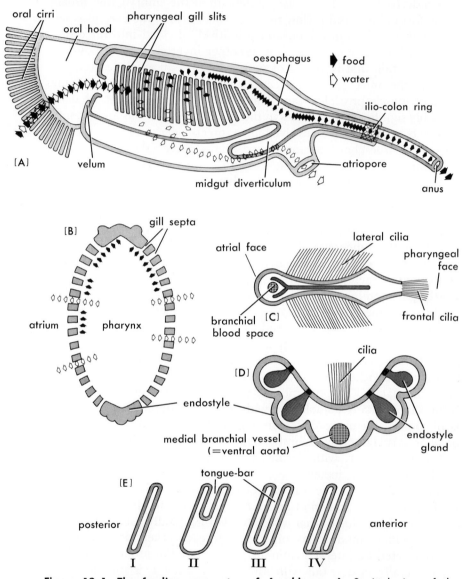

Figure 12·1. The feeding apparatus of Amphioxus. A: Sagittal view of the alimentary canal, and the outer tube of the atrium, with the major food and water currents. **B:** Cross section of the pharynx. **C:** Enlarged view of one gill septum in section. **D:** Enlarged view of endostyle in section. **E:** Four successive stages in the multiplication of gill slits.

12·1E). Anatomical features such as their skeletal supports, their ciliation and the blood vessels between, seem to be homologous throughout the chordates. Functionally, recent work has made it clear that they act as food-filtering organs, as in the bivalve molluscs, and this will be discussed in more detail below. Meanwhile, it is worth noting that the feeding function of the pharyngeal gill slits in the urochordates, hemichordates, and cephalochordates has been replaced by a respiratory function in most fishes and larval amphibians. However, reversion of function to filter-feeding is found in several distinct groups of fishes, including the largest living forms—basking sharks and whale sharks. The pharyngeal gill-clefts are of course only embryonic, transitory structures in the higher vertebrates, but the reader should remember that when equilibrating the pressure in his middle ears on change of altitude in an aircraft, he is utilizing the Eustachian tubes, the last functional vestiges of the food-filtering slits of the archetypic chordate. The dorsal hollow nerve cord is diagnostic, though only present to a limited extent in the hemichordates. The only other group of many-celled animals with nerves developed from the surface epithelium is the echinoderms. One functional consequence of this arrangement is that, as is well-known, several vertebrate neurosensory structures, including the vertebrate eye, develop centrifugally as outgrowths of the central nervous system. The inverted retina is a consequence. This curiosity of development, with the sensory ends of the receptor cells turned away from the source of sensation and toward the central nervous system, also occurs in some primitive chordates. Some textbooks suggest that this primitive dorsal hollow central nervous cord shows metameric segmentation. As we shall see in *Amphioxus,* this is not so in its primitive condition.

The coelom is of somewhat doubtful structural homology throughout the three protochordate groups, and it is certain that no *functional* homology throughout the chordates can be set up. In fact, it is only in the hemichordates that the coelom is a large body-cavity in the adult with any mechanical significance. The tripartite division of the coelom is seen in the pouches of the archenteron during the development of cephalochordates. It is an adult feature of the hemichordates where the head, collar, and trunk sections each contain a coelomic cavity.

The possession of a post-anal tail is another clearly chordate characteristic. Among the primitive chordates, only the Hemichordata have an anus which is terminal. In urochordate larvae, in cephalochordates, and in vertebrates, a muscular tail can be developed posterior to the opening of the alimentary canal. This tail typically consists of blocks of muscles on either side of the central axial skeleton, which as we noted above, is laterally flexible while being incompressible in length. The muscle segments of tailed chordates arise directly from the myotomes or mesodermal segments of the embryo. It is of some relevance in relation to theories of segmentation in chordates that in *Amphioxus* the myotomes are arranged alternately on either side of the axial notochord, *not* in the left and right pairs which would be expected with segmentation.

Other chordate characteristics concern the physiology of muscle action and ciliary patterns. The process of contraction in fast striated muscles involves the translation of chemical energy into mechanical energy by means still not finally established. However, ATP (adenosine triphosphate) is clearly involved, and ADP (diphosphate) reacts with the substances called phosphagens to produce ATP. Phosphagens, such as arginine phosphate and creatine phosphate, are thus reserves of energy-rich phosphate bonds. The main phosphagen found in invertebrates is arginine phosphate, and that characteristic of vertebrates is creatine phosphate. This was once thought to be a biochemical diagnostic feature of the chordates, with the echinoid echinoderms (which have creatine) and the hemichordates (which have both) being intermediate between the typical chordate animals and the invertebrates. However, the consistency of this biochemical characteristic proved unreal and based on ignorance of conditions in many other groups. Creatine phosphate—and other phosphagens—have been identified in certain annelid worms, in coelenterates, and in sponges. Although the absolute distinction no longer can be claimed, the principal phosphagen of most invertebrates remains that of arginine, and of vertebrates that of creatine.

Another physiological characteristic of the group was noted relatively recently. The cilia around the gill slits in the pharynx show metachronal waves, that is to say the cilia of one cell are slightly out of phase with those of the next and a wave appears to pass over the surface in the opposite direction to the effective stroke of the cilia. Viewed from outside the animal, the metachronal waves in the cilia of the pharyngeal gill slits seem always to run counterclockwise. That is, there is a functional asymmetry. (Think of this in relation to the wheels of a car or train viewed from outside.) This functional asymmetry of the gill slit cilia is found in *Amphioxus*, in several sea-squirts, and in some hemichordates. The developmental implications of this functional asymmetry are not yet clear: though they could imply derivation of the paired gill slits from a single series of openings, either ventrally or on one side. As was first pointed out by E. W. Knight-Jones and R. H. Millar, such functional asymmetry is unique among bilaterally symmetrical animals and seems to be a functionally homologous feature characteristic of the chordates.

Although the anatomy of *Amphioxus* is almost universally studied in courses on vertebrate morphology, aspects of its physiology, including its archetypic feeding mechanism, are less often documented. It is treated here as an invertebrate chordate. The anatomical features so important to comparative studies of vertebrates are neglected here; the features thought to be functionally homologous throughout the primitive chordates are, in contrast, emphasized.

Lancelets of the genus *Amphioxus* (sometimes known by the subgeneric name *Branchiostoma*) are not uncommon marine animals. They are found in shallow water over cleanish sand substrata, in restricted localities, but in all the world's oceans. They have been called "headless fish"; they are laterally flattened, spindle-shaped, about 2 inches long, and nearly translucent in life. Most of their life is spent half-buried, with the anterior end protruding above the surface of the sandy substratum. They can swim quite efficiently by fishlike, lateral flexures of the whole body. Both while semi-buried and while swimming, *Amphioxus* continues to feed by drawing a current of water in through the mouth and out via the pharyngeal gill clefts. Actually, the burrowing of *Amphioxus* is carried out by the same muscular movements used in swimming and really amounts to a vigorous swimming into the substratum. Ecologically, the sedentary feeding position taken up varies with the nature of the substratum: lancelets being more completely buried in coarse-grained sands but having most of the body (from oral hood to atriopore) protruding when in fine sand.

The great bulk of the tissues of the body are the muscles arranged on either side of the laterally flexible notochord. They form a series of units—which are *not* like metameric segments—termed myotomes, each V-shaped block being separated from the next by a myocomma of connective tissue and myotomes of each side alternate as can be clearly seen in cross sections of *Amphioxus*. The left-hand myotomes are the antagonists of those on the right, and swimming involves waves of contraction. The muscles of the anterior end contract first and are followed in regular order by muscles further and further back toward the tail along one side. The waves of contraction of the left-hand myotomes are out of phase with those of the right and these waves passing back along the animal create a forward motion through the water or sand. The central nervous system which controls this is of course a dorsal, hollow cord lying above the notochord. It has some enlargement anteriorly, with two so-called cranial nerves, and throughout most of its length, it gives rise to dorsal sensory roots and ventral motor roots running to the myotomes. Unlike the spinal cord and its paired roots in vertebrates, the sensory and motor nerves here alternate on either side of the nerve cord.

The mouth is protected by an oral hood with cirri and a velum. More than half the length of the gut behind it is the pharynx perforated with about one hundred gill slits on each side (see Figure 12·1A). There is a short midgut with midgut diverticulum and a hindgut with a ring of prominent cilia termed the ilio-colon ring. The anus is asymmetric on the left side of the ventral fin in front of the tail. In adult *Amphioxus* most of the gut is enclosed in an outer tube, the atrium, which opens through the atriopore (see Figure 12·1A). The filter-feeding system is basically simple: water and food particles pass in through the mouth, the water passing on through the gill slits into the atrium and out the atriopore while the food particles, entangled in mucus, pass posteriorly into the midgut section of the alimentary canal. The water current is created by the lateral cilia on the sides of the gill slits (Figure 12·1BC), all other cilia in the system being concerned with moving particles, or mucus. The details of the process show many analogies to filter-feeding in bivalve molluscs, and indeed were first worked out by J. H. Orton (one of the earliest workers on the functional morphology of molluscan feeding structures; see BLI), and further details of both feeding and digestion have been elucidated more recently by E. J. W. Barrington and Q. Bone. The feeding mechanism in *Amphioxus* derives its efficiency partly from the fact that, like the bivalves, there is considerable capacity for sorting particles, and rejecting, largely on a basis of size, those unsuitable for food. There is some preliminary sorting on the oral hood: larger particles being arrested on cilia of the hood tentacles and rejected by muscular flicking. The main water current created by the lateral cilia pulls water in through the mouth opening in the center of the velum, which water passes on through the atrium to the atriopore. Particulate material is collected on the frontal cilia and on mucus secreted largely by the glands of the endostyle. This last structure is formed by the imperforate central, ventral gutter of the pharynx (see Figure 12·1D), which is ciliated, contains four lines of endostyle glands, and is almost certainly homologous (both developmentally and functionally) with the thyroid of the vertebrates. The frontal cilia all beat in a dorsal direction and carry the captured food material and the mucus to the dorsal groove where it collects in a mucous cord. This is passed posteriorly by ciliary action to enter the digestive part of the alimentary canal. Digestion in *Amphioxus* is partly extracellular and partly intracellular. Enzymes are secreted by the midgut but principally by the diverticulum, and include protein-breaking and fat-breaking enzymes as well as amylase. The food cord passes slowly posteriorly, being relatively rapidly rotated by the cilia of the ilio-colon ring. Enzymatic globules secreted in the midgut diverticulum are carried by ciliary tracts to the main lumen of the gut where they impinge upon, and are incorporated into, the mucous rope. Small particles of partially digested food are broken off from the rotating rope and carried by other ciliary tracts into the dorsal side of the midgut diverticulum and onto its lateral walls. The cells here carry on intracellular digestion after taking up the fine particles. There is some absorption and consolidation of the mucous food cord in the hindgut and the faeces are discharged through the anus. The continuous rotation of the mucous rope thus involves trituration of the particulate food, the intimate mixing of food particles with digestive enzymes, and a sorting of partially digested material from more massive food. Thus it is almost exactly analogous in function to the rotation of the crystalline style in the bivalve gut (see BLI, pp. 144–148). In anatomically entirely different systems, cephalochordates and bivalves have found means of transmitting the effects of ciliary movement to the food material in the gut and achieving thereby trituration, intimate mixing, and sorting. Since most vertebrate textbooks refer to the midgut diverticulum of *Amphioxus* as the equivalent of the liver in vertebrates, it is worth pointing out here that functionally it has a closer equivalence to the pancreas of vertebrates, and an even closer functional similarity to the multiple digestive diverticula of filter-feeding molluscs.

In *Amphioxus,* the outer tube around the pharynx, the atrium, appears relatively late in development. It is structurally an invagination of the body wall which comes to surround the pharynx and to become the only other large internal space other than the gut: largely occluding the coelom in adult *Amphioxus.* Larval specimens of *Amphioxus* with about twelve pairs of gill slits live in the plankton and have a feeding mechanism exactly similar to that of adults, except that the water current which passes in the mouth to the pharynx passes out through the individual gill slits to the exterior directly. Taking an ecological view of the functioning of feeding mechanism, it is significant that the atrium, or outer tube, protecting the outer openings of the gill slits is developed at the time of so-called metamorphosis when the young *Amphioxus* settles to the bottom and exchanges its planktonic life for one involving burrowing in the substratum. This distinction seen in the "two ages" of *Amphioxus,* turns up again and again in many groups of primitive chordates. Thus, many primitive chordates living a planktonic existence have gill slits which open directly to the exterior, while those which burrow in the substratum have the outer openings of the slits protected in some way by structures which arise relatively late in development. If this ontogeny accurately reflects phylogeny, then atrial and other structures protecting the gill slits externally are in an evolutionary sense neomorphic.

Earlier circumstances in the development of the gill slits are also of considerable interest, and probably of evolutionary significance. When the embryonic gut first becomes functional, after about one hundred hours of development, the embryo is remarkably asymmetric. There is a very small mouth lying on the left side and a single gill slit behind it, considerably larger than the oral opening. Ciliary action moves water and food particles in through the mouth and out through the gill slit, and it seems that many particles pass straight through the system without being filtered. A few however are caught on the peripharyngeal band of cilia on the left side of the pharynx and these move past the slit and eventually back into the midgut. At this stage, the anus is nearly terminal but again asymmetric. The endostyle appears at an early stage but remains for some time on the right side of the pharynx as a gutter of secretory and ciliated epithelium. Further gill slits appear successively, but in two series. The primary series are ventral, initially, but then move up on the right side of the larva. Then the secondary series appears dorsal to the primaries again on the right side. After further growth, there is rearrangement with the primaries becoming the slits of the left side, the secondaries, those of the right side, and the endostyle gutter coming to lie between them. In the most studied species of *Amphioxus,* the larva at this stage has eight pairs of slits. The branchial slits then begin to multiply by the downgrowth of tongue bars (see Figure 12·1E) giving the opening of each slit as it divides a characteristic horseshoe appearance. This method of subdivision of gill slits is found in other primitive chordates.

A few other anatomical characteristics of cephalochordates should be mentioned. The coelom is reduced in adult *Amphioxus,* but has been claimed as showing tripartite division embryonically. Actually, at its greatest extent in the embryo, the coelom consists of five spaces: three anteriorly, two lateral and one dorsal; and two lateral trunk spaces which run back on either side of the gut (compare with Figure 12·3C). As already mentioned, these are squashed dorsally by the ingrowth of the atrium. The circulatory system is of immense interest to vertebrate zoologists, being almost part-for-part, structurally and functionally, homologous with the arterial and venous system of the most primitive fishes. There is no true heart, but a contractile ventral aorta from which afferent branchial vessels lead between the gill slits on the sides of the pharynx. The blood from these is carried by efferent vessels to paired dorsal aortae which unite behind the pharynx, the single posterior dorsal aorta giving rise to most of the arteries supplying the posterior gut and muscles. The blood returns to the posterior end of the ventral aorta (the "heart" position) through systems of cardinal veins and also through a "hepatic portal system," in a fishlike fashion. Functionally the direction of circulation, and the position of the respiratory organs, is clearly archetypic for the vertebrate pattern, blood passing anteriorly in the ventral blood vessels and posteriorly in the dorsal blood vessels. From a phyletic viewpoint, the excretory organs of *Amphioxus* are somewhat enigmatic. They consist of a series of protonephridia, closely similar to those found in some marine polychaetes, with bunches of solenocytes turning into short curved tubes opening into the atrium. There are about one hundred pairs of protonephridia associated with the occluded lateral coelomic cavities. The sexes are separate, and the gonads are developed on the atrial body wall. The gametes pass to the exterior through the atriopore. The small but yolky eggs are classic embryological material, with regular holoblastic cleavage.

The subphylum Cephalochordata consists of about twenty species, all very similar in form. In their simplest classification they all fall in one genus *Amphioxus*, the typical bottom-dwelling forms which have already been described falling in the subgenus *Branchiostoma*. The subgenus *Asymmetron* differs in having a single row of gonads and some of the characteristics of the so-called Amphioxides late larval form of *Amphioxus*. It seems probable that the species which show greater asymmetry, and the gill slits opening directly to the exterior, are forms which remain planktonic as adults.

Digression on Doubtful Allies

The only major group of primitive chordates showing any real ecological success is the subphylum Urochordata, the tunicates or sea-squirts. Before discussing them (Chapter 13) we must review a number of distinctly minor groups, of greater or less chordate affinities.

The minor phylum Hemichordata is divided into two unequal classes: the Enteropneusta for the numerous wormlike animals such as *Balanoglossus,* and the Pterobranchia for some rare but related forms which are sessile and have some resemblances to the lophophore-bearing minor phyla. The more typical enteropneusts, or acorn-worms, of such characteristic genera as *Balanoglossus* and *Saccoglossus* are wormlike animals which live a sedentary existence in semipermanent burrows on sandy shores between tide marks in temperate waters. Their burrows and sand-castings are remarkably similar to those of the lugworm, *Arenicola*. Their body is divided into three regions (see Figure 12·3A): a proboscis, a collar and a trunk region. The mouth opens behind the proboscis at the anterior end of the collar, branchial gill slits are represented by a series of pores in the anterior section of the trunk, and the anus is terminal. The general surface of the body is covered with a ciliated epidermis, in places forming ciliary tracts, of considerable importance in the way of life of the animals. There are also numerous mucous glands. The body wall is somewhat weakly muscularized, with no trace of mytomes, but with a tendency to an annelidlike arrangement

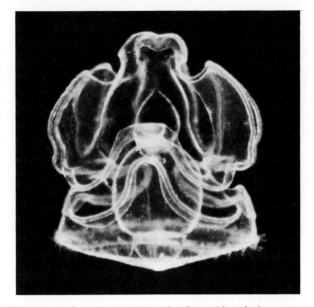

Figure 12·2. Tornaria larva of the hemichordate, *Glossobalanus*, in ventral view. Compare with Figure 11·5B and for further discussion, see text. [Photo © Douglas P. Wilson.]

of circular and longitudinal muscles. Perhaps related to their relatively simply organized body wall, most enteropneusts have great powers of regeneration.

The nervous system consists of a nerve plexus with certain definite tracts becoming consolidated into nerve trunks. It is peculiar in lying immediately under the ciliated epidermis and *above* its basement membrane. The primitive level of its organization is comparable only with those of the minor phylum Pogonophora, and the coelenterates; even the echinoderm system being more highly organized. In the trunk region of the body there are two consolidated tracts, one dorsal and one ventral, which are united anteriorly, that is in the boundary zone between trunk and collar, by a circular nerve. The circular nerve in the posterior end of the collar is also connected to the tubular dorsal nerve cord (the chordate diagnostic character) which runs only in the collar region. There is also another circular plexus on the surface of the proboscis stalk. Recent physiological and histological work suggests that the dorsal hollow nerve cord of the collar region seems to be essentially a through-conduction tract. There is no functional or microstructural evidence for major integrative activities. There are no obvious eyes or other highly organized sense-organs.

The enteropneust alimentary canal is a relatively simple tube from the mouth (posterior and ventral to the proboscis stalk) to the terminal anus. From the mouth an undifferentiated tube leads through the collar to the pharynx in the anterior part of the trunk. Various species vary in the number of gill slits (from twelve pairs to several hundreds) and they are usually incompletely divided by tongue bars (see Figure 12·4D). These gill slits do not open directly to the exterior but through gill-sacs and their gill-pores which are arranged in two laterodorsal lines on the surface. (see Figures 12·3A and 12·4A). The slits themselves have lateral cilia, as in *Amphioxus,* and are the agency which causes a water current to be drawn in continuously through the mouth and pass out through the gill-pores. In *Glosso-balanus* and *Ptychodera,* the pharynx is modified into an upper and a lower duct by lateral ingrowths (forming a figure-of-eight lumen in cross section; see Figure 12·4E). In contrast, in species of *Schizocardium* the pharyngeal condition is similar to that in the cephalochordates; there is no such division and only the dorsal groove and the endostyle are unperforated by the gill slits. There is an additional modi-

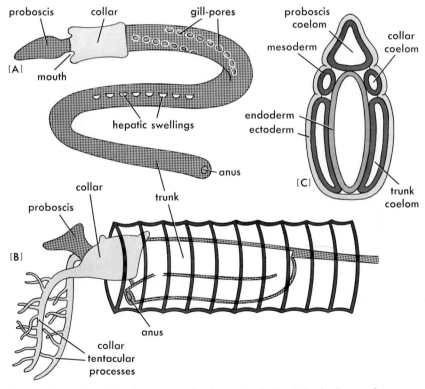

Figure 12·3. Hemichordate organization. A: Stylized body form of an enteropneust. **B:** Stylized body form of the minute sessile colonial pterobranch, *Rhabdopleura*. **C:** Late embryo of an enteropneust, showing the threefold (actually fivefold) division of the embryonic coelom.

fication in some species of *Ptychodera* and *Schizocardium* where latero-ventral fins of the body wall become greatly developed from the genital bulges or fins of most hemichordates. In these species genital wings are formed and enclose a false atrium. In a further development, in some species the gill-sacs may become reduced so that the U-shaped gill slits open from the pharynx directly into the space inside the genital wings. Behind the pharynx in the trunk there is an oesophagus which leads into a midgut extending for about a third of the length of the trunk. This midgut has blind sacs, the so-called "hepatic" sacs which show in some species as swellings on the outside of the body wall. There is then a short undifferentiated intestine to the terminal anus.

The feeding of enteropneusts involves three distinct mechanisms, of which the primitive ciliary filtration feeding of the other primitive chordates is probably the least important. The principal feeding mechanism is undoubtedly the collection of food particles by mucus on ciliary tracts of the proboscis and anterior edge of the collar. Our knowledge of this mucus-feeding mechanism results largely from the recent researches of E. J. W. Barrington, C. Burdon-Jones, and E. W. Knight-Jones. Since some misunderstandings have arisen concerning this feeding, it is worth noting that the ciliated surfaces of the trunk and most of the collar are *never* involved in the feeding mechanism. The third method of food collection involves the wormlike habit of engulfing sand and passing this through the alimentary canal for the extraction of suitable organic material.

The principal feeding mechanism involves the proboscis in secretion of mucus. The cilia upon it then carry the food particles in mucous threads backward and ventrally. These strands collect together (see Figure 12·4C) in the region of the preoral ciliary organ on the posterior face of the proboscis and becoming bound together into a mucous rope, pass from the underside of the proboscis stalk into the mouth. One of the few fast muscle reflexes found in these sessile worms is here involved in a rejection mechanism which allows the animal to interrupt the continuously acting food-gathering process. When larger particles, for example, big sand grains or distasteful substances come in contact with the proboscis base, the anterior edge of the collar reacts by moving forward closing the opening to the mouth and surrounding the base of the proboscis. After this has occurred the ciliary collecting tracts of the proboscis lead directly to the rejection tracts running posteriorly on the collar and trunk (see Figure 12·4B). The selectivity in ingested particles seems to result mostly from a bivalvelike sorting mechanism. No distinction is made except on mechanical grounds of size, but this gives rise to the rejection of larger sand grains, and the "acceptance" of the smaller animal and vegetable particles from the gathered particulate material. It has been suggested, however, that the ciliary preoral organ

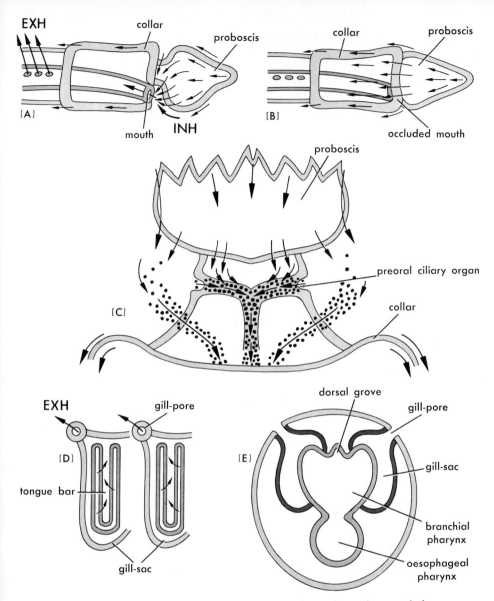

Figure 12·4. Ciliary feeding in enteropneusts. A: The surface ciliation of the anterior parts of an enteropneust while suspension feeding is going on. The water current into the mouth (INH) and out of the gill-pores (EXH) is created by the cilia on the side of the concealed pharyngeal gill slits. **B:** The rejection mechanism, in which withdrawal of the proboscis occludes the mouth, stops the pharyngeal water circulation, and aligns the ciliary tracks of the proboscis for rejection of particles posteriorly over the collar. **C:** The ciliary collection tracks of the proboscis and preoral ciliary organ of the proboscis neck, viewed ventrally. **D:** The gill-sacs which enclose the outer side of the U-shaped pharyngeal gill slits. **E:** Cross section of the pharynx in *Glossobalanus*, to show the arrangement of the gill-sacs and the almost divided lumen of the pharynx. [**C:** Adapted from C. Burdon-Jones in *Bol. Fac. F.C.L. Sao Paulo, 261, Zoologia,* 24:255–280, 1962.]

is capable of some tasting and that this results in the selection of particles on a qualitative basis quite apart from size.

The food material reaching the anterior part of the oesophagus is thus of three origins. In the most studied species, the most important source is the mucous rope deriving from the gathered strings of the proboscis, but there are also contributions from the filtering action of the gill slits in the usual chordate fashion and from material engulfed by swallowing substratum. It seems that the last component passes along the unperforated ventral part of the gut in those forms where the pharynx is subdivided (see above p. 185 and Figure 12·4E). From what knowledge we have of the variety of forms of hemichordates, it seems likely that the different proportions of these three types of food vary with ecological circumstances and perhaps correspond to the varieties of pharynx found. Behind the pharnyx, the wall of the midgut, and more especially the cells of the hepatic sacs, secrete the digestive enzymes in globules which are whirled by cilia into the continuous strand with the ingested food. The food itself never enters the hepatic sacs. The enzymes include amylase, protein-breaking and fat-breaking enzymes, and it is of interest that amylase is also secreted by the cells amid the ciliary mucus tracts of the proboscis so that, to some extent, digestion begins externally. Absorption takes place in the intestine and the faeces are consolidated before extrusion through the anus.

As a whole, the digestive system does not show any high level of specialization in structure or in function. There seems to be relatively little of the ciliary sorting mechanisms which are well-developed in other microphagous animals, for example, *Amphioxus* (see p. 180) and the bivalve molluscs (see BLI, pp. 144–148). However, longitudinal ciliated grooves have been described in the oesophagus and intestine of different species, and in different forms these may be single, asymmetric, paired dorsolateral, or dorsal and ventral. Further investigation may show that the specific patterns of these grooves, and their potential for particle sorting, and the variant patterns of gill-sacs and genital atria, do correspond to particular feeding habits and/or particular ecological niches.

It should be noted that the cilia and mucous glands of the proboscis, collar, and trunk also play an important role in burrow formation.

Some other aspects of structure and function in hemichordates are worth mentioning. In an embryonic *Balanoglossus* (see Figure 12·3C), the coelom has five divisions, one in the proboscis, two collar cavities, and two trunk cavities. These correspond to the three head and two body-cavities in *Amphioxus*, but in adult hemichordates they tend to become obscured by muscles and connective tissue. Excretion seems to be largely by the proboscis gland in the neck region, which gland surrounds the tissues once thought to be homologous with the notochord.

This region has a glomerulus of blood vessels connected to both the main dorsal blood vessel and the main ventral one. In adults, the two main vessels are both contractile and are responsible for all the circulation. The other blood vessels are mere crevices between tissues and organs. There is a so-called larval heart lying immediately posterior to the glomerulus on the dorsal blood vessel. There is some functional interest in that the blood circulation is annelidlike: that is, the propulsion is anterior in the dorsal blood vessel and posterior in the ventral blood vessel, with the afferent branchials leading dorsally from the ventral blood vessel and the efferents from the tongue bars of the gills into the dorsal blood vessel. In other words, the longitudinal circulation is in the opposite direction to that in *Amphioxus* and in fishes.

The sexes are separate with two rows of simple gonads in the body wall of the anterior part of the trunk. As already mentioned, these may develop into genital wings. There are no gonoducts or other accessory structures and each gonad opens separately to the exterior. Fertilization is usually external, but there is considerable variability in the type of development. In several species, there is complete segmentation leading to a pelagic larva termed a tornaria (see Figure 12·2), which has close resemblances to the auricularia larva of holothurians, though it has an additional perianal band of cilia. Unlike the echinoderms, however, there is no drastic metamorphosis, the tornaria growing into a little tripartite larva with proboscis, trunk, and collar. Thus bilateral symmetry is retained from the larva into the adult. The pulsating vesicle, corresponding to the madreporite of echinoderms, becomes the rudimentary heart of the young adult. In other species, there is a more direct development with the hatching embryo being already wormlike and never passing through a planktonic phase as a ciliated larva. This is the case in *Saccoglossus kowalevskii*, which is the only common hemichordate of the Atlantic coast around Cape Cod. Some evolutionary zoologists attach great importance to the tornaria larva as evidence of chordate-echinoderm affinities. This has perhaps been overstressed, as in other cases where larval similarities have been used as a basis for phylogenies (see Chapters 9 and 11).

The remaining hemichordate class, the Pterobranchia, involves a few species placed in two genera, *Cephalodiscus* and *Rhabdopleura,* both minute sessile colonial forms, dredged from deep water of the oceans, and relatively rare, though worldwide in distribution. Although the individual zooids in colonies of both genera show the three divisions of the hemichordate body—proboscis, collar, and trunk—the trunk is reflexed on itself so that the gut is U-shaped. The collar in *Rhabdopleura* has two arms (see Figure 12·3B), and in *Cephalodiscus* many pinnate ones. Feeding is presumably by the ciliation on these and on other parts of the anterior surface, but the only account of a living pterobranch dates from 1915, and contains several ambiguities and unexplained complications. Further study of the patterns of ciliation and of feeding is long overdue, and with increased numbers of deep-sea research vessels, the coincidence of living pterobranchs and suitable investigators may occur soon. *Cephalodiscus* has one pair of pharyngeal gill slits, and *Rhabdopleura* has a pair of probably homologous dorsolateral grooves in the pharyngeal wall. Of course, if one disregards possible feeding significance, there can be no need for respiratory circulation in animals so small (*Cephalodiscus:* 1–5 millimeters; *Rhabdopleura:* 0.1–1.0 millimeter). Both genera have a single pair of gonads in the anterior part of the trunk, and thus there is no replication of gill slits or of gonads or of muscles in these primitive chordates. Asexual budding is prevalent and probably is responsible for colony formation. Sexual development has not been well worked out, but late larval forms are known with a pronounced resemblance to young enteropneusts, that is having a wormlike tripartite body with a terminal anus.

There are some other, even more doubtful, "chordate allies." The minor phylum Pogonophora consists of wormlike tube-dwellers, again living in the deepest waters of the oceans. The group has claims to be the most recently discovered phylum of animals—the twentieth-century phylum—since they were first dredged early this century and thoroughly investigated only about twenty years ago. At least forty-five species are now known, mainly described by Russian zoologists. They have a long, thin, tripartite body with the anterior part bearing a single tentacle or many tentacles, and they have some other deuterostome characters including an enterocoelic body-cavity with tripartite division and a pair of coelomoducts. They have a closed circulatory system with a heart, but remarkably for animals about 20 centimeters long, no trace whatsoever of an alimentary canal. Details of feeding and digestion remain obscure. The chitinous tubes which they secrete have some resemblance to those of *Rhabdopleura,* and it is to the fact that the eggs of some species are brooded within the tube that we owe our knowledge of the holoblastic cleavage and enterocoelous body-cavity formation.

Early Chordate Fossils

The wholly extinct phylum Graptolithina has skeletal organization which resembles the tubes secreted by *Rhabdopleura,* and although the great bulk of fossil graptolites show no traces of internal structures, there are a few well-preserved petrifactions for which some other pterobranch characteristics are claimed. (Some paleontologists, however, relate this relatively extensive fossil group to the phylum Cnidaria.)

Certain other groups of Palaeozoic fossils are more clearly chordate and, in addition to "experimental" stocks, include the forms considered to be the earliest-known vertebrates. Perhaps the most significant of these from our viewpoint, though not the earliest in time, is the genus *Jamoytius,* discovered and described from Silurian deposits by Erroll I. White about twenty years ago. It was jawless, and had simple myotomes like *Amphioxus,* lateral fin folds, a persistent notochord, and no bony internal skeleton. The chordate stocks from which vertebrates are derived must have included similar forms. Armored, jawless fish (sometimes collectively referred to as ostracoderms) are found in Silurian and Devonian deposits, with fragmentary remains in the Ordovician. Among these are the cephalaspids which have been meticulously reconstructed (initially by E. H. O. Stensiö and his Scandinavian colleagues), and show a truly vertebrate organization of skeleton, brain, and sense-organs. Their pharyngeal apparatus conforms to the primitive chordate pattern we have described as archetypic and it is almost certain that they fed by stirring up bottom deposits, sucking in organic detritus through the jawless mouth and filtering it on the ten pairs of gill slits. Among one stock of such agnathous fishes, biting jaws were later evolved, and thence came the bulk of vertebrates.

It is worth noting here that at least two workers consider the fossil

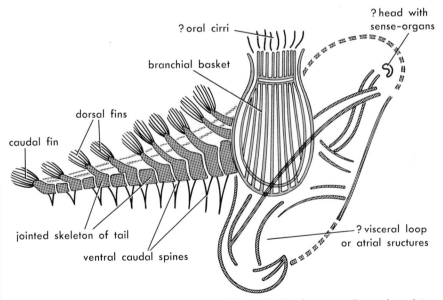

Figure 12·5. An enigmatic "unassigned" fossil chordate, usually neglected in phylogenies, a diagrammatic reconstruction of *Ainiktozoan loganense* from Silurian rocks in Scotland. This curious assemblage of sea-squirt and vertebrate characteristics is discussed in the text. [Largely based on D. J. Scourfield in *Proc. Roy. Soc. London* (B), *121*:533–547, 1937; and original specimens in the Hunterian Museum of the University of Glasgow.]

carpoids as primitive chordates, basing this on supposed branchial slits, and even more doubtful homologies with the cranial skeleton and neural cavities of primitive vertebrates. (In this book, the carpoids were treated as primitive echinoderms; see Table 10·1, and pp. 146–147.) That controversial allocation to these two distinct phyla is possible, could be taken as highly suggestive. An almost equally bizarre, but more certainly "chordate" fossil is the "unassigned" species *Ainiktozoon loganense* from Silurian rocks in Scotland. Over thirty specimens have been found and they are all similar, consisting of a "branchial basket" like that of a sea-squirt with an attached, jointed, and finned tail like those of craniate chordates. A reconstruction is shown in Figure 12·5, and it is immediately obvious that although probably chordate, this organism could not be assigned to any subphylum as they are presently diagnosed. *Ainiktozoon* was first described in detail by the late D. J. Scourfield, a distinguished worker on crustaceans, and this eponymously enigmatic fossil organism seems to have been largely ignored by zoologists constructing phylogenies of the chordates. Of course, there are many other enigmatic fossils. Although *Ainiktozoon* is included in this book partly because I have handled most of the extant specimens (in the Hunterian Museum of the University of Glasgow) and been disproportionately impressed by them, the species has real evolutionary significance. The possible interrelationships of the chordates will be considered later (Chapter 14), but the assemblage of characteristics in this fossil form suggests that there were several "experimental" chordate stocks in the Palaeozoic, from which there survive the minor but important group of cephalochordates, and the two widely divergent "successful" lines—the sea-squirts (Urochordata) and the vertebrates.

Invertebrate Chordates II:
The Successful Sea-squirts

IF THE VERTEBRATES ARE EXCLUDED, the subphylum Urochordata includes the most successful animals built on the chordate ground plan.
They are successful by all our usual measures: there being over two
thousand species, which make up a considerable part of the animal biomass in some marine environments. On suitable rocky substrata in the
lower littoral and sublittoral, sessile tunicates (solitary or colonial) are
among the commonest marine invertebrates. Less commonly, and for
limited periods of time, pelagic urochordates (both solitary and colonial) can be the dominant animal organisms of the marine plankton.
The great majority of species are allocated to the class Ascidiacea, the
true sea-squirts, or ascidians. There are two other classes of urochordates: the Larvacea and the Thaliacea, less numerous groups, both
highly specialized for permanent life in the marine plankton. Solitary
forms of typical sea-squirts (class Ascidiacea) will be discussed first.

Functional Organization

Solitary ascidians are cylindrical or spherical animals of moderate size
(rarely over 10 centimeters long), attached at one end to rocks or man-
made hard substrata like pilings, dock walls, and ships' bottoms, and
bearing at the other two openings— oral and atrial. The genus *Ciona*
(see Figures 13·1A and 13·2A) can be regarded as showing an almost
archetypic simplicity of structure and function. The protective integument, termed the tunic, or test, is a cuticular secretion of the ectoderm
but has in it wandering mesodermal cells and blood vessels. It was
formerly thought unique among animals because the ground substance,
tunicin, includes cellulose as a constituent, though this supporting ma-

terial, typical of plants, has now been found to occur in the connective tissues of the dermis of mammals. The muscles lie in the body wall beneath the test in the form of longitudinal and circular strands. There are also sphincter muscles around the openings. There is no real skeleton, the muscles partly being antagonized by the elasticity of the test and partly working around an enclosed hydraulic skeleton of the seawater in the pharyngeal and atrial spaces. There is a plexus of nerves on the body wall and this is attached to a single central ganglion, lying between the oral and atrial openings on the side of the pharynx (see Figure 13·2C). There are no complex sense-organs, but mechanoreceptors are well-developed around the oral and atrial openings. Some experimental work has suggested chemoreception, and there can be pigment spots which have been termed ocelli although there is no physiological evidence of photoreception.

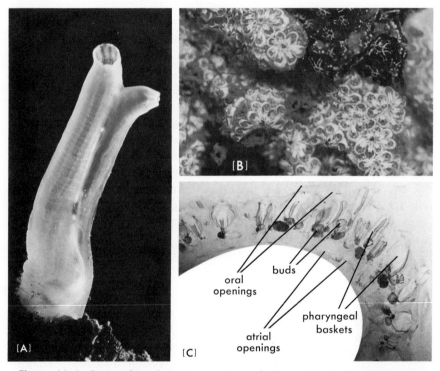

Figure 13·1. Sea-squirts. A: Living specimen of *Ciona intestinalis,* a solitary sea-squirt. **B:** Living specimens of two color varieties of the colonial sea-squirt *Botryllus schlosseri,* growing on a mooring line. Each "flower" system consists of a group of seven or eight zooids with their separate oral openings surrounding a single common atrial opening. **C:** Part of a cross section through the pelagic colonial form, *Pyrosoma.* In the wall of the tubular colony the oral openings of the individual zooids face outward, the atrial openings toward the internal "lumen" of the colony. [**A:** Photo © Douglas P. Wilson. **B** and **C:** Photos by the author.]

Internally, the pharynx is proportionately enormous, with the gill slits subdivided, forming an elaborate basketwork. The atrium forms an outer bag laterally and dorsally around the pharynx, and the anus and genital openings lie within it. In life, when a sea-squirt is expanded, water is continuously passing in the oral opening to the pharynx, through the multiplied gill slits to the atrial cavity, and out through the atrial opening. The oral opening leads into a short, wide buccal cavity and thence into the prebranchial zone of the pharynx. This is bounded posteriorly by the peripharyngeal band, two ciliated ridges which lead to a dorsal tubercle on the atrial side of the pharynx. Here there is a ciliated funnel leading through a duct to a gland below the ganglion—the subneural gland. All these structures are relatively well-innervated and a sensory function has been proposed. Apart from this sampling of the branchial current, in a fashion analogous to the osphradium of molluscs, other functions have been proposed for the subneural gland. In some forms, it is supposed to secrete the hormone which induces spawning. In others, including *Ciona,* it is known to be a site of phagocytic action, dead blood cells being removed from it and passed out the duct into the pharynx and the alimentary system. The ascidian subneural gland has been homologized with the preoral ciliary organ of enteropneusts, and with the preoral pit of the developing *Amphioxus,* and thence—though the homology may be latent at best—with the pituitary gland of vertebrates. (This last of course has a twofold origin: the adenohypophysis is a preoral invagination and the neurohypophysis is a neural downgrowth from the floor of the brain.) Behind the peripharyngeal band and subneural gland lies the enormous branchial basket. All the bars are hollow, with blood lacunae running through them, and are covered with a ciliated epithelium which does not show much differentiation. However, the cilia of the sides of the openings (the stigmata) are somewhat longer, could be termed lateral cilia, and cause the main water current through the sea-squirt. There is a ventral imperforate tract forming an endostyle, homologous with that in *Amphioxus,* and there is also the dorsal hyperpharyngeal band from which, in *Ciona,* hang a row of processes termed the languets. (In some other genera there is a continuous dorsal lamina.) The pharynx leads through a small oesophagus to the stomach, thence to the intestine, and through the rectum to the anus in the atrium (see Figure 13·2A).

Appropriately enough, the processes of feeding and digestion in sea-squirts have been best worked out by investigators also concerned with filter-feeding molluscs (including J. H. Orton, C. M. Yonge, and R. H. Millar). Unlike the filtration in bivalves (see BLI), feeding in sea-squirts is dependent on the secretion of a mucous net. This is produced by the mucous gland cells of the endostyle which secrete on to

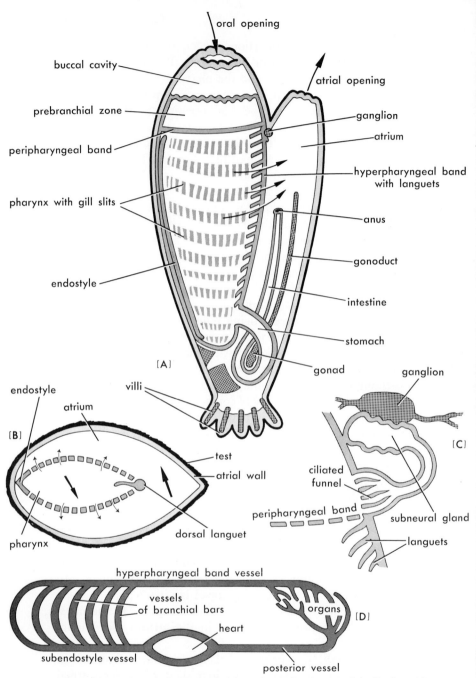

Figure 13·2. Sea-squirt organization. A: Stylized vertical half of a solitary sea-squirt showing the feeding water current through the enlarged pharyngeal basket. **B:** Cross section through the pharynx and atrium. **C:** The ganglion and subneural organ. **D:** Diagrammatic circulatory system of sea-squirt. Direction of circulation can be clockwise or counterclockwise. For further discussion, see text.

the inner face of the basketwork, the cilia of which beat dorsally, thus pulling a sheet or net of mucus across to the dorsal row of languets. Meanwhile, a continuous water current is drawn in through the oral opening, down through the pharynx, and moved through the stigmata into the atrium by the lateral cilia. Food particles picked up on the mucous net are carried dorsally. On the row of languets, the mucus is worked posteriorly as a cord or rope into the oesophagus. The oesophagus itself is ciliated and the stomach is the main region of enzyme secretion. Some ascidians, though not *Ciona,* have secretory diverticula in this region. The cilia of the grooves of the stomach are arranged in spiral rifling so that the food is churned with the enzymes in a mucus-bound mass. Absorption takes place in the midgut and the remaining material is carried out to the anus by ciliary action, since there is no peristalsis. The filtration in Urochordates can be fantastically efficient: the mucous net in some forms being capable of retaining particles of between 1 and 2 microns in diameter. In contrast to *Amphioxus,* to the enteropneusts, and indeed to most ciliary mucus-feeders, the sea-squirts have little in the way of rejection mechanisms for disposing of large particles. Ecologically, this is connected with their life on harder substrata, filtering relatively clean seawater.

Although as chordates, the sea-squirts are classified as coelomate animals, there is no definite evidence of any true coelom existing in adult urochordates. The small epicardial sacs which lie beside the heart in some forms, and which are "enterocoelic" in that they communicate with the posterior end of the pharynx, are thought to be vestiges of a once more extensive coelom. The evidence, however, is not conclusive, and functionally the epicardium is concerned in excretion in some sea-squirts, and is part of the stolonic asexual budding system in others. The circulatory system, and the blood within it, are equally peculiar. The blood consists of a colorless plasma with various corpuscles, some nucleated. It may contain fantastically high concentrations of unusual elements, such as vanadium and niobium. (The concentration of vanadium in the sea is of the order of .0002 milligrams per liter, while the concentration in the plasma of *Ciona* is 400 milligrams per liter, and in some other ascidian species where it is contained in corpuscles reaches concentrations of 1.8 grams per liter.) The smaller ducts of the circulatory system really lack walls and are essentially the spaces between organs. The heart at the base of the pharynx is a muscularized fold in the pericardial wall giving rise to a vessel running anteriorly under the endostyle, and another posteriorly to the digestive organs and gonads directly. The vessels in the branchial basket arise from the vessel under the endostyle and unite on the other side of the pharynx in a hyperpharyngeal band vessel which sends branches to the digestive organs, gonads, etc. Structurally, this is diagrammed in Figure 13·2D, but functionally the contractions of the ascidian heart are peristaltic for about ten contractions driving blood in one direction, then, after a pause, contractions start again, but in the opposite direction. Thus alternately, the blood system resembles that of *Amphioxus* and fishes (i.e., when it leaves the heart by the subendostyle vessel and returns by the posterior vessel), and alternately, that of *Balanoglossus* and annelid worms when it leaves the heart by the posterior vessel and returns by the subendostyle vessel. It was earlier claimed that the regular reversals of direction of heart beat were due to the jamming of corpuscles in the branchial bar vessels, causing a back-pressure, which in turn caused reversal of the peristaltic action of the cardiac musculature. This process certainly occurs, but the isolated heart has since been found to reverse its beat. In itself this reversal would be evidence against back-pressure as a control, but M. E. Kriebel has now demonstrated that the two pacemakers controlling the contractions are themselves affected by changes in intracardiac pressure and can show adaptation.

The Ascidian Tadpole

Many tunicates, including *Ciona,* are hermaphroditic: all mature animals have both a compact ovary and a ramifying testis within the loop of the alimentary canal in the base. The two genital ducts have their external openings within the atrium, but fertilization—normally cross-fertilization—takes place in the sea. In the most typical cases, the small egg undergoes a regular segmentation, developing to a larva rather like that of *Amphioxus.* In Figure 13·3A is illustrated the peculiar appendicularian larva or ascidian tadpole. Both the embryonic development and the free-swimming tadpole larval stage in ascidians are relatively brief. There are also many forms where the free-swimming larval stage is lacking, being retained during development within the maternal atrium, and in some cases, showing little or no development of the tail. However, in those forms, presumably archetypic, where the free-swimming tadpole is fully developed, it exhibits the full integrated concert of structures and functions which we regard as diagnostic of chordates. Dorsal to the gut with its pharynx and gill slits lies a hollow, central nervous system, and there is a propulsive tail with, as an axial skeleton, a typical notochord. Unlike *Amphioxus,* the tail muscles are not arranged in myotomes.

After a varying period of larval life, the ascidian tadpole attaches by the adhesive papillae lying below the mouth to a suitable substratum for adult life, and almost immediately the processes of metamorphosis begin. There are two processes of change which occur simultaneously:

the tail is reduced, and the internal organs of the body undergo a rotation of 180° (see Figures 13·3A, B, C). Most accounts suggest that the tail is absorbed by phagocytosis. This is incorrect; the method, recently investigated by Richard A. Cloney and by James W. Lash, is a controlled shrinking of the outer epidermis which pulls the notochord and muscles into the body where reorganization takes place. The rotation brings the oral opening (see Figure 13·3C) to a position opposite the attachment papillae. N. J. Berrill has presented evidence that sug-

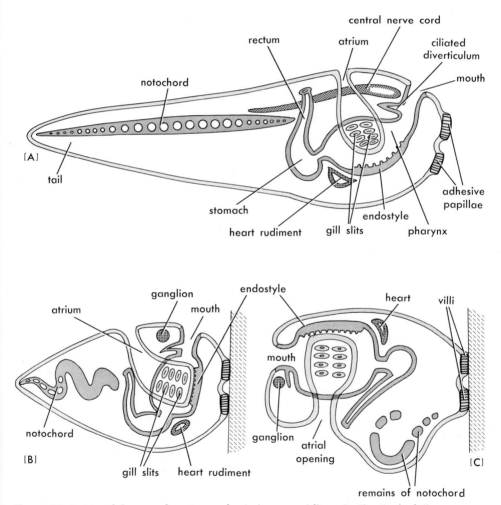

Figure 13·3. Larval form and metamorphosis in an ascidian. A: The "tadpole" or appendicularian larva, whose locomotory tail is lost at metamorphosis. **B** and **C:** Successive stages in the process of metamorphosis which follows attachment to a suitable substratum: a controlled shrinking of the outer epidermis pulls in the tail, and the internal organs undergo a rotation of 180°.

gests that the tail epidermis reacts to buildup of metabolites after the muscular activity of dispersal. However, the manner in which rotation of the organs, due to increased growth between the mouth and the adhesive papillae, and the reduction of the tail, occur simultaneously immediately after successful settlement, suggests some kind of overall regulatory control.

Sea-squirt Diversity

The class Ascidiacea is a large group of animals, encompassing considerable diversity of form and of habitat. They are surprisingly difficult to subclassify since they exhibit very complex interrelationships, in many cases involving considerable convergence of structural patterns. They are all sessile animals but show various degrees of colonial habit, associated with various types of budding. The so-called simple ascidians can be divided into two main groups using diagnostically what seem to be anatomical characters of some phyletic significance. There is a *Ciona* group with dorsal languets, a perivisceral cavity developed from the epicardial pouches, and having associated compound forms, mostly arising from stolonial budding. In contrast, the *Ascidia* group show a continuous dorsal lamina in the pharynx, no epicardial pouches, and have their associated compound forms mostly resulting from pallial budding, that is, budding of the so-called mantle, or pallium—the outer body wall.

Three principal types of budding are found: two stolonial and one pallial. In the most complex stolons a double tube arrangement is found, with a median ventral outgrowth of the visceral region forming the outer tube with an inner tube derived from the epicardium and with some mesenchyme cells lying between the two tubes. A second simpler form of stolonial budding involves an outer tube with no inner one, but merely a strip of mesoderm cells. Pallial budding usually involves an outgrowth of the atrial wall, each bud having a lining corresponding to the atrial lining and an outer coating of the outer body wall and its integument. It is worth noting that the concept of three determinate cell layers in triploblastic animals breaks down in the processes of this budding. For example, in the stolonial budding of forms like *Aplidium,* the parental endoderm of the epicardial tube forms not only the alimentary canal in the bud, but also the ectodermal central nervous system and atrium, and the mesodermal gonad. In the pallial budding of Botryllidae, the parental ectoderm of the atrial wall forms not only the central nervous system in the bud, but also the endodermal alimentary canal, and the mesodermal gonad. In the budding and regeneration processes of most other groups of many-celled animals, the germ layers are usually deterministic: mesodermal structures being derived from mesoderm and so on. However, other exceptions to the concept include ectoproct budding and holothurian regeneration.

In ascidians, all degrees of colonialism—from loose associations to highly organized interdependence—are found. Apart from the solitary forms which separate completely immediately after budding, there are the so-called social ascidians, like *Clavelina* and *Perophora,* where the only link is a common stalk, and *Polyclinum,* where the individual zooids are completely separated but are embedded in a common test which is well-developed and massive. Perhaps the commonest colonial forms are those like *Botryllus* and *Didemnum* which form spreading patch colonies. As shown in Figure 13·1B, the individual zooids produced by budding—the blastozooids—are arranged in flowerlike systems, each group of seven or so sharing a common cloaca or united atrial opening. Forms like botryllids grow all over suitable substrata as rather indefinitely organized patches. In contrast, forms like *Distaplia* show a more highly organized colonial habit, with budding in the colony limited to one region so that the colony as a whole has a definite shape and shows some division of function between groups of zooids.

It seems clear that the various levels of colonial habit have evolved several times in the ascidians, so that the old-fashioned classification into simple ascidians and composite ascidians is unreal. However if one tries to base a classification of the ascidians on the basis of the three types of budding which occur, then it quickly becomes apparent that this will not match with the basic structural features which provided a dichotomy between the *Ciona* group and the *Ascidia* group. Many phylogenies and "natural" classifications of the sessile tunicates have been proposed. None is entirely satisfactory. As a whole, the diagnosis of the class Ascidiacea would run: "tunicates where the tailless adult is sessile, with the mouth and atrial opening subterminal at the same end." This sufficiently distinguishes the largest group of the urochordates from the remaining planktonic forms.

Four Patterns of Planktonic Life

Perhaps the most fantastic way of life found in primitive chordates is that of the urochordate class Larvacea, diagnosed as tunicates in which the sexually mature animal retains the organization of the larva including a locomotory tail with a notochord. The "head" of the tadpole, propelled by the tail, is like a much simplified sea-squirt, with a simple pharynx with two gill slits and no atrial cavity. Both gill slits and anus open directly to the exterior. Species of such genera as *Oikopleura, Megalocercus,* and *Fritillaria* occur in the marine plankton in many parts of the world. The most bizarre aspect of their biology is that they build an external house of hardened mucous secretions. As shown in Figure 13·4A, this house is used as a filter-feeding apparatus. Continuously repeated flexures of the tail cause a water current to be drawn in through the coarse filter and thence through the internal net. Particles trapped on the internal net are ingested. Internal pressure opens the flap valve at the "posterior" end of the house, and the jet drives the

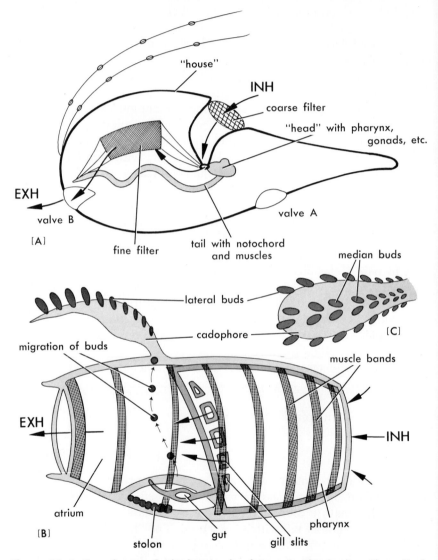

Figure 13·4. Two forms of planktonic chordates. A: The feeding "house" of *Oikopleura*, a larvacean urochordate. For full description, see text. **B:** The adult asexual oözooid of the thaliacean urochordate, *Doliolum*, with its characteristic barrel-shaped body and muscle bands. Note the blastozooid buds migrating to the cadophore. Details of the complex life-cycle, with its alternation of sexual and asexual reproduction, are given in the text. **C:** The cadophore of *Doliolum* viewed from above.

animal forward. When coarse particles accumulate on the outer filter, the animal discards its house, escaping through the spring-loaded trap door at the anterior end and constructing an entirely new house in about an hour. In a few species, including those of *Fritillaria,* the adult tadpole lives outside its secreted house during the feeding process. Larvacean tunicates live their entire life in the plankton, and have a simple life-cycle involving only sexual reproduction. No form of budding occurs in the group.

It is generally agreed that the Larvacea are urochordates which have become specialized by the retention of an essentially larval organization after the onset of maturity and the development of gonads. Thus, they are one of the classic examples of the evolutionary process which is termed neoteny. The hypothesis implicit in the concept of neoteny is that during the evolution of a stock of animals, the developmental processes and life-cycle have been so modified that sexual maturity becomes attained in association with larval organization. Like so many other brilliant—but essentially intuitive—hypotheses regarding larvae, the concept that neoteny has occurred several times in the evolution of chordates was first developed by Walter Garstang (see, for example, the discussion of larval torsion in gastropods, in BLI pp. 124–126).

The third and last class of the urochordates is the Thaliacea, diagnosed as tunicates where the tailless adult is pelagic with the mouth and atrial openings at opposite ends. In these oceanic forms, the water current through the body is not only used for feeding and respiration but also as a means of locomotion. The class splits into three rather distinct orders: Pyrosomatida, Salpida, and Doliolida. The first order, with its single genus *Pyrosoma,* consists of tropical species in pelagic colonies, with structure which is intermediate between that of typical sea-squirts (Ascidiacea) and those of the salps and doliolids.

Certain points of functional organization are distinctive of the class Thaliacea, apart from the planktonic habit and positions of the openings, and these must first be discussed. There is budding which is basically stolonial but the complex stolon may contain elements of different cords of cells: from the atrium, from the gonad, from nervous tissues. Individuals which develop from fertilized eggs—oözooids—differ greatly from the asexually-produced blastozooids. In all three groups the oözooids have lost the power of sexual reproduction. In salps and doliolids the blastozooids have lost the power of budding. Thus, in these two groups there is a regular alternation of generations. Throughout the Thaliacea, there is a reduction in the importance of cilia, water currents being assisted in *Pyrosoma* by muscular movements and caused by them in salps and doliolids. These groups could be considered as analogous to the Cephalopoda and Septibranchia among the molluscs (see BLI) where muscles have replaced cilia in creating the water currents through the mantle-cavity. Particularly in doliolids, the muscle strands form strong complete rings encircling the barrel-shaped body and by their contraction, create the water current which serves, not only for feeding but for jet propulsion.

In *Pyrosoma* there is no larva and the oözoid is retained within the parental atrium where it produces the first four buds of the blastozooid generation. These are freed and the oözooid regenerates while the four primordial buds form the typical cylindrical colony by further budding (Figure 13·1C). The colonies are not uncommon in tropical waters and are brilliantly luminescent, responding readily to tactile stimulation. The list of oceanographers who have written their names "in letters of fire" on the sides of newly captured *Pyrosoma* colonies begins with Wyville Thomson, Murray, Moseley, and company on the *Challenger* Expedition, and has not yet ended.

In the order Salpida, there is again no larva, but the oözooid leads a separate existence as an asexual form, with a coiled stolon from which blastozooids are regularly budded. These blastozooids, which are sexual individuals unable to bud, are shed in relatively short chains, the form of which differs in different species. There is thus a regular alternation of generations. Salps are among the few primitive chordates important in the economy of the sea. In the plankton, they both compete with and prey on young fish and other elements of the zooplankton. They appear irregularly, but in enormous numbers, in certain areas of the North Atlantic, and the years when salps occur represent very poor years for young fish. Some investigators have attempted to correlate variations in the occurrence of salps with the sunspot cycle.

Doliolum shows the most complex life-cycle of thaliaceans, indeed of all chordates. The solitary sexually mature adult, which is a blastozooid, produces three eggs, each of which develops to a tadpole larva. Each larva metamorphoses to the adult asexual oözooid in which there is a stolon which continuously buds off blastozooids. These blastozooid buds (see Figure 13·4BC) migrate through the atrial wall to a structure called a cadophore where further budding results in the formation of groups of polymorphic blastozooids forming a colony. There are usually two lateral lines of gastrozooids or feeding individuals, and a median row of phorozooids or nurses, and gonozooids or sexual individuals. Eventually, solitary blastozooids which are sexually mature adults break away and give rise to the eggs again. Unlike the case in the salps, the solitary sexual blastozooid does not differ much in gross anatomy from the adult asexual oözooid.

It is almost certain that these four groups of planktonic urochordates have evolved from more typical sessile ascidians, with a tadpole larva. The Larvacea have probably evolved by neoteny. On the other hand, in *Pyrosoma* and in the salps, the tadpole larval stage has been suppressed. As N. J. Berrill has pointed out, the ascidian larval tail could hardly propel the large, barrel-shaped body of a thaliacean, and as a consequence, the powerful muscle bands have been evolved from the circular muscle strands which are present in all adult ascidians. However, the limited capabilities of the ascidian tail may have been surpassed in at least one extinct organism (see *Ainiktozoon,* p. 192). Certain other aspects of the phylogeny of chordates are discussed in Chapters 12 and 14.

CHAPTER 14

A Partial Phylogeny of Invertebrates

AMONG THE DEFINITIONS given by *Webster* for the adjective "partial" are: "relating to the part rather than the whole," and "biased." Both are applicable to this short chapter.

As stressed in the first chapter, and elsewhere in this book and in BLI, all statements regarding homologies, archetypes, and phylogenies involve hypotheses. Hopefully, there is matter in the two books other than such theory, and most of it concerns whole animals discussed at the mechanistic-physiological and adaptive-functional levels of explanation. In other words, the reader can turn (with Louis Agassiz and Wordsworth) "to the solid ground of Nature" and confirm or correct any statements describing the mechanics of feeding in *Artemia* by careful observations on living brine-shrimps. Thus also can be verified or modified any discussions of restorative regeneration in sponges or in polychaetes; of ciliary sorting in bivalves, in sabellid worms, or in *Amphioxus;* of suspension-filtering by sponges, by gastropods, by barnacles, by lampshells, or by sea-squirts; of locomotory forces and their application in ribbon-worms, in leeches, in brine-shrimps, in millipedes, in starfish, or by the tails of chordates. But this is not all of biology, even after restricting consideration to the organismic grade of complexity (and thus ruling out almost all concern with ecosystems on one hand, and with "molecular" biology on the other). Even such a miniature attempt at a pandect of invertebrate biology as these two volumes has involved the use of systematics and phylogeny as linking material. In the study of biology at any level, both the arbitrary distinctions of higher taxonomic categories, and the inductive concepts of homology in organ systems or in functional processes, are of pragmatic value.

Such theoretical matters *do* aid human comprehension of animal diversity. Their implicit biological significance, however, cannot readily be tested. There can be no verification by observation of the living animal, nor any appropriate application of "the scientific method" in a crucial experiment. Essentially this is because the hypotheses involved in archetypes, homologies, and phylogenetic classifications all concern past time more distant than Proust's *"temps perdu."* No matter how extensive and how convincing the evidence for any statement on the phylogeny of an animal stock, that statement remains merely a "reasonable inference." It can never be turned (lacking a time-machine) into a notarized genealogy. Similarly, such animal phyla as the Mollusca and Echinodermata must each have had *real* ancestral progenitors. Although the synthesis of studies in comparative anatomy, functional morphology, paleontology, and embryology allows us to set up more or less convincing functioning archetypes for these two phyla, such hypothetical types are not ancestors.

The above may seem to involve some overemphasis on a simple distinction between the verifiable facts of structure and function and the hypotheses of common descent. However, any acquaintance with the literature of any branch of biology can demonstrate that biologists—from ecologists to biochemists—have need to be periodically reminded of this distinction, and hence of the speculative and biased nature of all phylogenies.

Homologies and Phylogenies

The specific bias of the phylogenies presented here depends on a belief that natural selection has promoted functional interdependence so that a stereotyped pattern of whole animal physiology has emerged in each major group of animals (see Chapter 1). It follows that perception of functional homologies within a group—along with the widely accepted homologies of structure—can lead to the concept of an archetype as a working efficient machine with such a pattern of functional integration.

After these disclosures of its subjective and biased nature, the purpose of this chapter is the construction and defense of Figure 14·1, as a rough phyletic "tree." This involves the possible interrelationships between the thirty phyla of many-celled animals set out in Table 1·1. The perceptive reader will have noted that a preliminary question has thus been bypassed. The grouping of the million or so "species" of many-celled animals into these thirty phyla has itself involved acceptance of ideal arbitrary categories. Defense of each of these phyla as reflecting some objective reality of common descent and interrelationship has been set out in the chapters of BLI and this book, in some cases, mashalling extensive and convincing evidence of "common descent," and in other cases, otherwise. Once the thirty phyla are ac-

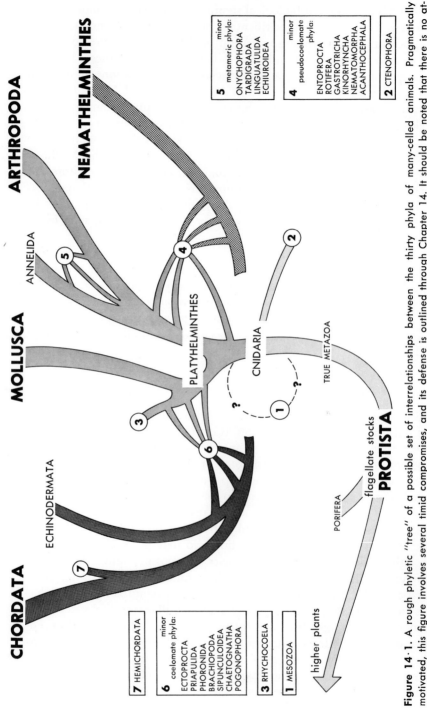

Figure 14·1. A rough phyletic "tree" of a possible set of interrelationships between the thirty phyla of many-celled animals. Pragmatically motivated, this figure involves several timid compromises, and its defense is outlined through Chapter 14. It should be noted that there is no attempt to represent *time* but only to depict possible degrees of interrelationship.

CHORDATA

ARTHROPODA

NEMATHELMINTHES

MOLLUSCA

ANNELIDA

ECHINODERMATA

PLATYHELMINTHES

CNIDARIA

TRUE METAZOA

PORIFERA

flagellate stocks

PROTISTA

higher plants

5 minor metameric phyla:
ONYCHOPHORA
TARDIGRADA
LINGUATULIDA
ECHIUROIDEA

4 minor pseudocoelomate phyla:
ENTOPROCTA
ROTIFERA
GASTROTRICHA
KINORHYNCHA
NEMATOMORPHA
ACANTHOCEPHALA

2 CTENOPHORA

7 HEMICHORDATA

6 minor coelomate phyla:
ECTOPROCTA
PRIAPULIDA
PHORONIDA
BRACHIOPODA
SIPUNCULOIDEA
CHAETOGNATHA
POGONOPHORA

3 RHYNCHOCOELA

1 MESOZOA

cepted, then the despairing evolutionary biologist might be tempted to draw, instead of a "tree," thirty separate diverging lines originating in a blank "unknown." Moved by pragmatic needs, or by the seduction of intellectual exercises in matters ultimately unprovable, many biologists venture further. Some would regard Figure 14·1 as a timid and retrograde step in the continued evolution of phylogenies.

Most biologists accept that many-celled animals evolved from protistans—whether, as in the majority view, from flagellate stocks or, as in a vocal minority opinion, from multinucleate protozoans, need not concern us here. It seems clear that the Parazoa (phylum Porifera, the sponges; see BLI p. 64) evolved independently of the rest of the Metazoa. The relationships of the phylum Mesozoa remain obscure. The present phylogeny regards the rest of many-celled animals as having a common origin. In this stock, the phylum Cnidaria is clearly the most simply constructed of many-celled animals and is here regarded as encompassing the most primitive. This does not necessarily imply that the other triploblastic phyla were derived from forms like presently living adult coelenterates. The probable phylogeny within the cnidarians is discussed in BLI (pp. 51–52), and it can be argued that the first coelenterate could have been a ciliated, nematocyst-bearing organism resembling a simplified version of the actinula larva found in some hydrozoans. The ctenophores are obviously allied with the cnidarian coelenterates (see BLI p. 56), though it is impossible to derive the ctenophores from any of the present stocks.

The simplest triploblastic animals are free-living flatworms belonging to the order Acoela in the phylum Platyhelminthes. It is generally supposed that such flatworms were derived from planuloid ancestors, that is, motile, gutless, larval coelenterates with bilateral symmetry. A contrary hypothesis, not accepted here, is associated with the name of J. Hadži, and suggests that the coelenterates are not primitive metazoans but are secondarily derived from the flatworms. Apart from other difficulties, this implies that the most primitive cnidarians were forms like sea-anemones, a phylogeny of coelenterates not readily acceptable (see BLI).

The ribbon-worms of the phylum Rhynchocoela are obviously closely allied to the flatworms, and can be regarded as comprising several experimental patterns of worm organization. Ignoring for the moment the various groups of minor phyla, three or possibly four stocks of triploblastic animals with body cavities originate from the flatworm stock. The phylum Nemathelminthes stands apart from all other stocks as the only successful group of pseudocoelomate worms. Nematodes, though enormously successful, show no obvious relationship to any of the other phyla (see BLI, Chapter 7). There is a group of six minor phyla, the rotifers being the most successful, which are built on the pseudocoelomate plan. They have often been grouped together in various ways (see BLI), but the case for a superphylum like Aschelminthes is not a good one. The minor pseudocoelomate phyla are best regarded as all separate stocks, some perhaps distantly allied to the Nemathelminthes.

The case for regarding the major phylum Mollusca as being derived from the turbellarian-rhynchocoel phyla, directly, is set out in Chapter 14 of BLI. It hinges on the conclusion that metameric segmentation (as it occurs in annelids and arthropods) does not occur in primitive molluscs. The discovery of *Neopilina* provoked a reconsideration of ideas on the phylogeny of the molluscs and their relationships with other phyla. A number of zoologists have used it as a link between the arthropod-annelid stock and the rest of the molluscs, a major premise being that several features in both *Neopilina* and the chitons are primitive metameric characters. However, the evidence against the alleged metamerism of chitons is very strong, and probably, despite *Neopilina*, annelidlike segmentation has never occurred in an animal which could be called a mollusc. It should be noted that the basic molluscan plan of structure and function is remarkably uniform throughout the group, and the homologies established, particularly as regards the mantle-cavity and the gills, are extensive and convincing.

Apart from metamerism, there are undoubtedly some features in which the annelid-arthropod stock shows a pattern of specialization from the platyhelminth-rhynchocoel stem-group which paralleled that of the molluscs. The difficulties of this are bypassed in Figure 14·1. The case for a close relationship, though distant in time, between the annelids and the arthropods has already been set out (see pp. 8–20). Annelid metamerism must have arisen in rhynchocoel-like worms which became coelomate. By the development of a metamerically divided body-cavity, which could be used as a hydrostatic skeleton, annelids became the most successful group of wormlike animals. The question of the origin of the arthropods in some primitive stock of annelidlike animals has already been discussed (see pp. 122–123). The possibility that the arthropods were of multiple origin in protoannelid stocks was there considered and dismissed. The diverse stocks of arthropods, including the enormously successful crustaceans, arachnids and insects, *do* show a uniquely integrated pattern of organs and functions and can be best considered as a single monophyletic group. There are four minor phyla (see Chapter 8) more or less closely connected to the annelid-arthropod stock. Given our ignorance of their true relationships, it is best to treat the Onychophora, the tardigrades, the linguatulids, and the echiuroid worms as separate minor phyla, though clearly metameric.

There remain eight minor coelomate phyla, of which two are probably connected with the stock of the echinoderms and chordates. The remainder are discussed in Chapter 9, and as with the pseudocoelomate minor phyla, interrelationships are obscure. As already discussed, the common possession of a lophophore as a food-collecting organ may or may not be phyletically significant. The phylum Ectoprocta is a moderately successful group, as was the phylum Brachiopoda at an earlier period of the geological time-scale.

In some ways the phylum Echinodermata stands as distinct as the nematode worms from the other groups of many-celled animals. However, as has been discussed (see pp. 172–173), their origins and those of the chordates can be postulated as close. The fact that some fossils, described by some investigators as primitive echinoderms, can be regarded by others as chordates, is itself significant (see p. 192). It seems likely that there were several "experimental" chordate stocks in the Palaeozoic, from which three stocks survive. Apart from the minor group of *Amphioxus* and its allies, the successful chordate groups are the sea-squirts and the vertebrates.

As has been often discussed, the probably neotenous origin of the Larvacea among the urochordates is significant in relation to the origin of true vertebrates. Whatever the exact historic sequence, it is obvious that the first vertebrates possessed many of the features which we have attributed to the chordate archetype. Both the excellent fossil record and several physiological characteristics confirm that a group of agnathous fishes—the ostracoderms—comprised the most primitive truly vertebrate animals. The beautiful reconstructions of ostracoderms by Stensiö and his associates have convinced most zoologists that these armored jawless fishes used a filter-feeding mechanism involving the pharyngeal gill slits, as does *Amphioxus* and larval sea-squirts on the one hand and the young stages of the presently-living parasitic cyclostomes on the other. In addition, there are several cases of structural and functional homologies which link the vertebrates to the primitive chordate stocks. There is anatomical, developmental, and biochemical evidence for homologous origins of the endostyle of sea-squirts and the thyroid gland of vertebrates. Several others are mentioned in Chapters 12 and 13. In fact, to conclude this discussion of Figure 14·1 and invertebrate phylogeny in general, it can be said that the evidence linking the vertebrates with the invertebrate chordates is more extensive and convincing than that which can be used to link any two of the phyla of invertebrates to each other.

Envoi

Some recapitulation may be allowed to a Scots pedagogue. Discussion of structural and functional homologies is of value in trying to comprehend the prodigious diversity of form and function in invertebrates. So also is utilization of the working archetype, in which the synthesized concert of structures and functions forms an integrated functional plan, as a basis for phylogeny. However, all inferred phylogenies at all levels involve hypotheses. *Real,* but unknowable, ancestral progenitors are mirrored, with variable distortion, in our hypothetical archetypes. *Actual,* but untraceable, genealogies are reflected—less or more efficiently—in our circumstantially determined classification. Once more, archetypes are *not* ancestors.

Further Reading

A COMPLETE SET of references is impossible within this size of book. There is a vast and steadily increasing literature on invertebrate animals. Accordingly, only brief annotated citations of four classes of works follow. These are general textbooks (mostly modern, one-volume works), comprehensive surveys (mostly older, and largely anatomical), accounts of specific groups, and modern surveys of invertebrate physiology. Almost all of the last three classes contain extensive bibliographies, but any student reading up on a topic in depth should learn to use *Biological Abstracts* and *Zoological Record* in the preparation of his own reference list.

Two good, relatively up-to-date, general texts on the invertebrates are R. D. Barnes, *Invertebrate Zoology* (Philadelphia: Saunders, second edition, 1968) and P. A. Meglitsch, *Invertebrate Zoology* (New York: Oxford University Press, 1967). Among the older textbooks which remain valuable (mostly quoted in revised editions) are R. Buchsbaum, *Animals Without Backbones* (Chicago: University of Chicago Press, 1948); F. A. Brown, Jr., *Selected Invertebrate Types* (New York: Wiley, 1950); R. W. Hegner, *Invertebrate Zoology* (New York: Macmillan, 1933); L. A. Borrodaile, F. A. Potts, L. E. S. Eastham, J. T. Saunders (and G. A. Kerkut), *The Invertebrata* (Cambridge: Cambridge University Press, 1961); and G. S. Carter, *General Zoology of the Invertebrates* (London: Sidgwick and Jackson, 1951).

The finest modern comprehensive survey in English is, unfortunately, incomplete. Libbie H. Hyman's *The Invertebrates* (New York: McGraw-Hill, 1940 and subsequent dates) includes two volumes on groups dealt with here (echinoderms and most of the minor coelomate groups), but does not yet cover annelids or arthropods. In French, the excellent series *Traité de Zoologie* (Paris: Masson et Cie., 1948 and subsequent dates) is edited by P.-P. Grassé, and though still incomplete, covers several of our groups. In German, a series more useful in some sections than others, *Klassen und Ordnungen des Tierreichs* (Leipzig: Friedlander und Sohn, 1873 and subsequent dates to present) was originally edited by H. G. Bronn. Another useful work, comprehensive and complete, although dated, is the *Handbuch der Zoologie* (Berlin: Walter de Gruyter, 1923), which was edited by W. Kukenthal and T. Krumbach. Two older series in English are still valuable: E. Ray Lankester's *Treatise on Zoology* (London: Adam and Charles Black, 1900–1909) and the *Cambridge Natural History* (London and New York: Macmillan, 1895–1909). Finally, two modern compilations prepared for other purposes include fine up-to-date surveys of some invertebrate groups: *Treatise on Invertebrate Paleontology,* edited by R. C. Moore (Lawrence, Kansas: Geological Society of America, 1952 and subsequent dates to present), and *Treatise on Marine Ecology and Paleoecology,* edited by J. W. Hedgepeth (New York: Geological Society of America, 1957).

Modern accounts of specific groups of "higher" invertebrates are not extensive. E. J. W. Barrington's *The Biology of Hemichordata and Protochordata* (Edinburgh: Oliver and Boyd, 1965) is excellent of its kind. R. Phillip Dale's *Annelids* (London: Hutchinson, 1963), J. Green's *A Biology of Crustacea* (London: H. F. and G. Witherby Ltd., 1961), J. L. Cloudsley-Thompson's *Spiders, Scorpions, Centipedes, and Mites* (Oxford: Pergamon, 1958), and D. Nichols' *Echinoderms* (London: Hutchinson, 1962) are all useful. *Echinoderm Biology,* edited by N. Millott (New York and London: Academic Press, 1967) contains several valuable reviews and bibliographies. Certain older books remain valuable, and these include the *Handbook of the Echinoderms of the British Isles* by Th. Mortensen (London: Humphrey Milford, 1927), *The Arachnida* by T. H. Savory (London: Edward Arnold, 1935), *A Textbook of Arthropod Anatomy* by R. E. Snodgrass (Ithaca: Cornell University Press, 1952), and *A General Textbook of Entomology* by A. D. Imms, O. W. Richards, and R. G. Davies (London: Methuen, 1957). Dealing with a single genus—but one of immense importance (see pp. 69–73) and in superb fashion—is *The Biology of a Marine Copepod* by S. M. Marshall and A. P. Orr (Edinburgh: Oliver and Boyd, 1955).

Lastly, there are a number of relevant modern surveys of invertebrate physiology, of which *Comparative Animal Physiology,* by C. L. Prosser and F. A. Brown, Jr. (Philadelphia: Saunders, 1961) is perhaps the finest. E. J. W. Barrington's *Invertebrate Structure and Function* (Boston: Houghton Mifflin, 1967) covers some aspects of comparative physiology excellently and concisely. In German, W. von Buddenbrock's six-volume *Vergleichende Physiologie* (Basel: Birkhauser, 1953) is most useful. Two valuable short books for students are J. A. Ramsay's *Physiological Approach to the Lower Animals* (Cambridge: Cambridge University Press, 1952) and E. Baldwin's *An Introduction to Comparative Biochemistry* (Cambridge: Cambridge University Press, 1948). In its more circumscribed scope, T. Bullock and A. Horridge's *The Structure and Function of the Nervous System in Invertebrates* (San Francisco and London: Freeman, 1965) is extensive, detailed, and excellent. Similarly, C. M. Yonge's *The Sea Shore* (London: Collins, 1949), A. C. Hardy's *The Open Sea* (Boston: Houghton Mifflin, 1956), and J. A. C. Nicol's *The Biology of Marine Animals* (New York: Pitman, 1960) provide excellent accounts of certain aspects of physiological ecology in invertebrates. A detailed synthesis of existing knowledge on one aspect of invertebrate nutrition is provided in the *Biology of Suspension Feeders* by C. Barker Jørgensen (Oxford, Pergamon, 1966). Surveying the coelom and locomotory mechanics in invertebrates, and discussing evolution with this particular bias, is the excellent *Dynamics in Metazoan Evolution* by R. B. Clark (Oxford: Clarendon Press, 1964). For more circumscribed groups, the two volumes of *Physiology of Crustacea,* edited by T. H. Waterman (New York: Academic Press, 1960 and 1961), and the *Physiology of Echinodermata,* edited by R. A. Boolootian (New York: Interscience, 1966), contain several useful reviews and extensive bibliographies.

Index